Experimental design for the life sciences

Experimental design for the life sciences

SECOND EDITION

Graeme D. Ruxton
University of Glasgow

Nick Colegrave
University of Edinburgh

OXFORD
UNIVERSITY PRESS

OXFORD
UNIVERSITY PRESS

Great Clarendon Street, Oxford OX2 6DP

Oxford University Press is a department of the University of Oxford.
It furthers the University's objective of excellence in research, scholarship,
and education by publishing worldwide in

Oxford New York

Auckland Cape Town Dar es Salaam Hong Kong Karachi
Kuala Lumpur Madrid Melbourne Mexico City Nairobi
New Delhi Shanghai Taipei Toronto

With offices in

Argentina Austria Brazil Chile Czech Republic France Greece
Guatemala Hungary Italy Japan Poland Portugal Singapore
South Korea Switzerland Thailand Turkey Ukraine Vietnam

Published in the United States
by Oxford University Press Inc., New York

First published 2003

British Library Cataloguing in Publication Data
Data available

Library of Congress Cataloging in Publication Data
Data available

Typeset by Newgen Imaging Systems (P) Ltd., Chennai, India
Printed in Great Britain
on acid-free paper by
Antony Rowe Ltd, Chippenham, Wiltshire

ISBN 0–19–928511–X 978–0–19–928511–2 (Pbk)

3 5 7 9 10 8 6 4 2

To Hazel and Becky

Preface

How to read this book

This book is an introduction to experimental design. We mean it to be a good starter if you have never thought about experimental design before, and a good tune-up if you feel the need to take design more seriously. It does not come close to being the last word on experimental design. We cover few areas of design exhaustively, and some areas not at all. We use the Bibliography to recommend some good books that would facilitate a deeper consideration of design issues. That said, it is also important to realize that the basic ideas of design covered in this book are enough to carry out a wide range of scientific investigations. Many scientists forge a very successful career using experiments that never take them outside the confines of the material covered in this book. This book will also help you tackle more advanced texts, but if you absorb all that we discuss here, then you may find that you know all the experimental design that you feel you need to know.

This book is about how to design experiments so as to collect good quality data. Of course, that data will almost certainly need statistical analysis in order to answer the research questions that you are interested in. In order to keep the size of this book manageable, we do not enter into the details of statistical analysis. Fortunately, there are a huge number of books on statistical analysis available: we even like some of them! We recommend some of our favourites in the Bibliography. We also provide some pointers in the text to the types of statistical tests that different designs are likely to lead to.

We often illustrate our points through use of examples. Indeed, in some cases we have found that the only way to discuss some issues was through examples. In other cases, our point can be stated in general terms. Here we still use examples to amplify and illustrate. Although we think that you should aim to read the book right through like a novel at least once, we have tried to organize the text to make dipping into the book easy too.

Our first job will be to remove any doubt from your mind that experimental design is important, and so we tackle this in Chapter 1. Chapter 2 discusses how good designs flow naturally from clearly stated scientific questions. Almost all experiments involve studying a sample and extrapolating conclusions about that sample more widely; how to select a good sample is the key theme of Chapter 3. The nitty-gritty of some simple designs forms the basis of Chapter 4. There is no point in formally designing an elegant experiment if you then make a poor job of actually collecting data from your sample individuals, so Chapter 5 contains some tips on taking effective measurements. Chapter 6 is a compendium of slightly more specialized points, which do not fit naturally into the other chapters but at least some of which ought to be useful to your branch of the life sciences.

This is *not* a maths book. Nowhere in this book will you find the equation for the normal distribution, or any other equation for that matter. Experimental design is not a subsection of maths—you don't need to be a mathematician to understand simple but effective designs, and you certainly don't need to know maths to understand this book.

Good experimental design is vital to good science. It is generally nothing like as difficult as some would have you believe: you can go a long way with just the few simple guidelines covered in this book. Perhaps most amazingly of all, it is possible to derive enjoyment from thinking about experimental design: why else would we have wanted to write this book!

On the second edition

The coverage has not changed dramatically from the first edition, although all sections have been rewritten—sometimes extensively so—for greater clarity. We have also attended to filling in a few gaps, most notably in substantially increasing our consideration of the special challenges associated with using human subjects. However, the big change in this edition is the way in which the material is presented. As well as extending the use of some of the features of the previous edition, we also include a number of new features that we hope will substantially increase the ease of use of this book. We introduce these new features on the following two pages.

Learning features

Key definitions

There is a lot of jargon associated with experimental design and statistical analysis. We have not tried to avoid this. Indeed, we have deliberately tried to introduce you to as much of the jargon as we can. By being exposed to this jargon, we hope that it will become second nature to you to use—and understand—it. This should also make reading more advanced texts and the primary literature less daunting. However, to make negotiating the minefield of jargon more straightforward, we have increased the number of definitions of key terms provided and increased the level of detail in all definitions. Each key word or phrase is emboldened at the point where we first use it, and is given a clear definition in the margin nearby.

3.4 Randomization

Randomization simply means drawing random samples for study from the wider population of all the possible individuals that could be in your sample.

An easy way to avoid many sources of ps
that your experiment is properly randomiz
simplest techniques to use in experimen
the most misunderstood and abused. *Pro*
any individual experimental subject has
individual of finding itself in each expe
randomization can avoid many sources of

Statistics boxes

To emphasize the important link between good experimental design and statistics, we now include a number of statistical boxes. These boxes should help you to see how thinking about design helps you think about statistics and vice versa. We have added the boxes at points where we think that keeping the statistics in mind is particularly useful and include pointers in the main text to each box.

STATISTICS BOX 2.3 **Generalizing from a study**

An important issue in the interpretation of a study is how widely applicable the results of the study are likely to be. That is, how safely can you **generalize** from the particular set of subjects used in your study to the wider world? Imagine that we measured representative samples of UK adult males and females and found that males in our sample were on average 5 cm taller. Neither we nor anyone else is particularly interested in our specific individuals; rather, we are interested in what we can conclude more widely from study of the sample. This becomes easy if we think about the population from which our sample was drawn. Our aim was to create a representative sample of UK adults. If we have done this well, then we should feel confident about generalizing our results to this wider population. Can we generalize even wider than this: say to European adults or human adults in general? We would urge a great deal of caution here. If your

Self-test questions

The more you think about experimental design, the easier designing robust experiments becomes. To get you thinking while you are reading this book, we now include a number of self-test questions in every chapter. Often there will not be a clear right or wrong answer to a question, but suggested answers to all questions can be found at the back of the book.

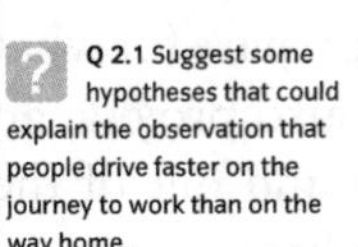

hospital ward respond more effectively
other end of the ward. Candidate hypothe

1. Patients are not distributed randoml
   having a tendency to place more sever
   room away from the entrance.
2. Patients nearer the door are seen first o
   getting more of the doctor's time and a
3. Patients respond positively to increas
   contact associated with being nearer t

Take-home messages

To help to consolidate your thinking, we end most sections with a take-home message. By including these throughout the text we hope to give you an immediate opportunity to evaluate your understanding of a section before moving on.

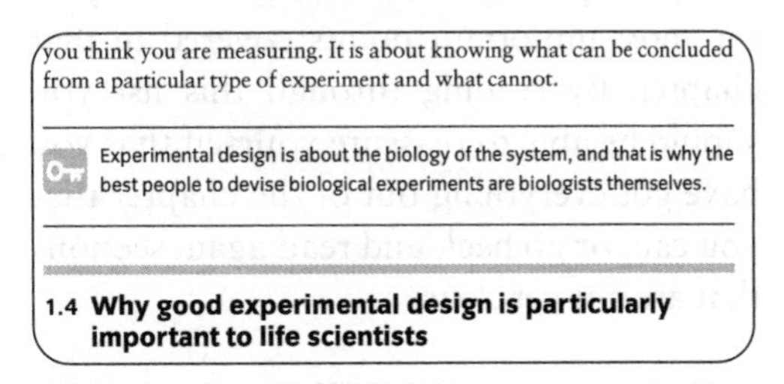

Boxes

We have aimed to produce a book that can be read from cover to cover, and so have tried to keep unnecessary details out of the main text. However, in areas where we feel that more details or more examples will lead to a fuller understanding of a concept, we have included supplementary boxes.

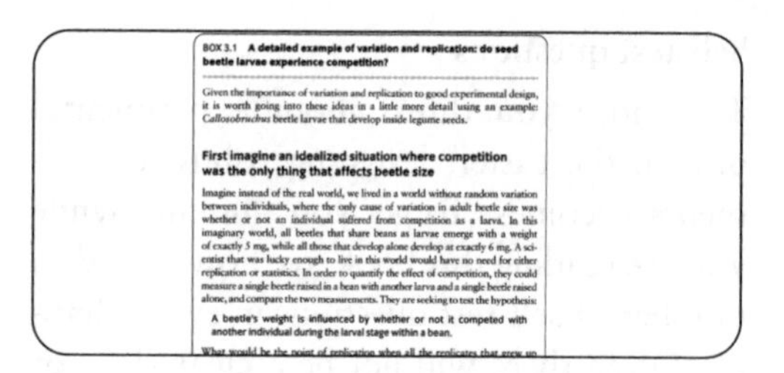

BOX 3.1 **A detailed example of variation and replication: do seed beetle larvae experience competition?**

Given the importance of variation and replication to good experimental design, it is worth going into these ideas in a little more detail using an example: *Callosobruchus* beetle larvae that develop inside legume seeds.

First imagine an idealized situation where competition was the only thing that affects beetle size

Imagine instead of the real world, we lived in a world without random variation between individuals, where the only cause of variation in adult beetle size was whether or not an individual suffered from competition as a larva. In this imaginary world, all beetles that share beans as larvae emerge with a weight of exactly 5 mg, while all those that develop alone develop at exactly 6 mg. A scientist that was lucky enough to live in this world would have no need for either replication or statistics. In order to quantify the effect of competition, they could measure a single beetle raised in a bean with another larva and a single beetle raised alone, and compare the two measurements. They are seeking to test the hypothesis:

A beetle's weight is influenced by whether or not it competed with another individual during the larval stage within a bean.

Chapter outlines

Every chapter begins with an outline of the main points covered by that chapter. These outlines should help you to prepare for what is ahead. The outlines will also provide an easy way for you to dip in and out of the book during subsequent readings.

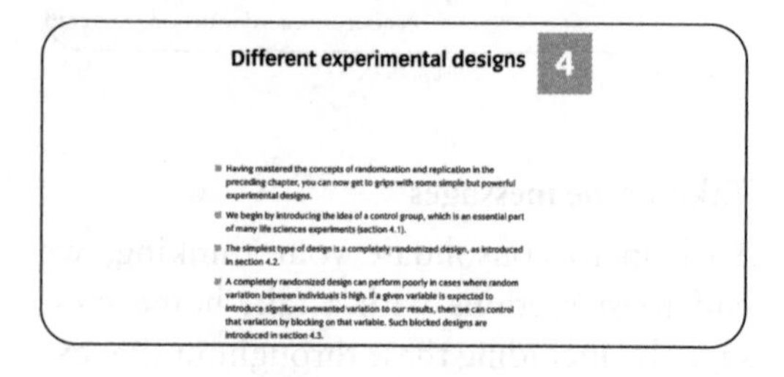

Different experimental designs 4

- Having mastered the concepts of randomization and replication in the preceding chapter, you can now get to grips with some simple but powerful experimental designs.
- We begin by introducing the idea of a control group, which is an essential part of many life sciences experiments (section 4.1).
- The simplest type of design is a completely randomized design, as introduced in section 4.2.
- A completely randomized design can perform poorly in cases where random variation between individuals is high. If a given variable is expected to introduce significant unwanted variation to our results, then we can control that variation by blocking on that variable. Such blocked designs are introduced in section 4.3.

Chapter summaries

Every chapter finishes with a summary of the most important points covered in that chapter. By reading through this list you should be able to reassure yourself that you have got everything out of the chapter that you can, or go back and read again sections that are not yet clear.

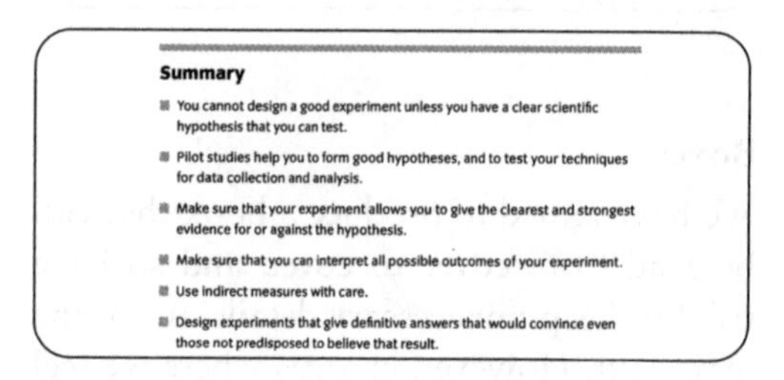

Summary

- You cannot design a good experiment unless you have a clear scientific hypothesis that you can test.
- Pilot studies help you to form good hypotheses, and to test your techniques for data collection and analysis.
- Make sure that your experiment allows you to give the clearest and strongest evidence for or against the hypothesis.
- Make sure that you can interpret all possible outcomes of your experiment.
- Use indirect measures with care.
- Design experiments that give definitive answers that would convince even those not predisposed to believe that result.

Ethical issues

We cannot overemphasize the importance of ethical issues when designing studies involving living organisms (including humans). In this new edition we highlight sections of particular relevance to ethics by placing a symbol in the margin at the appropriate point.

orates? The only way to know is to try it, and this will ultimatel
maximum size of the experiment you can plan for and carry out.
There is a lot to be said for practising all the techniques and
that you will use in your experiment before you use them in y
experiment. If your experiment will involve dissecting mosqu
want to make sure that you can do this efficiently, so as to be sur
will not destroy valuable data points and cause unnecessary suff
Trial runs of different types of assays can also allow you to fine-
so that they will give you useful results. You might be planning a
for female butterflies in which you will give a female a choice of se

Flow chart

The exact process of designing an experiment will vary considerably between studies. Nevertheless, there are key stages in the design process that will apply to most if not all studies. The flow chart at the end of this book is intended to summarize and guide you through the main stages of designing an experiment. We have indicated at each point in the chart the sections of the book that are most relevant.

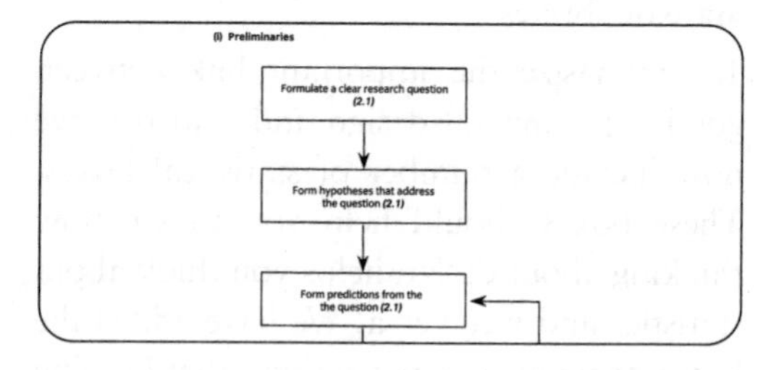

Acknowledgements from the first edition

Graeme has been moaning at the representatives of scientific publishers for years that there was not a book like this available. However, it was Sarah Shannon of Blackwells that suggested that he stop moaning and write one himself. It was also Sarah who suggested that Nick would be an ideal co-author. We may have ended up at another publisher, but we are both very grateful to Sarah for her support and advice at the inception of this book.

At OUP, Jonathan Crowe has been everything we could have hoped for in an editor. He commented on two drafts, and found just the right combination of criticism and enthusiasm to spur us on. He also found excellent reviewers (Roland Hall, Keith McGuiness and two others), who in their own diverse ways, have also really helped to improve the finished work. Also at OUP, John Grandidge and Emily Cooke were models of friendly efficiency. Many students from the 2001–2002 BSc Zoology and Aquatic Bioscience courses at Glasgow, and the Zoo4 QZ course at Edinburgh, provided useful feedback. In particular, Sandie King provided particularly full and useful comments. Fiona McPhie read the whole thing and gave honest feedback on some bad jokes that we've thankfully removed.

The text was carefully proof-read by Hazel Ruxton and the late David Ruxton, who managed to remove an embarrassing number of errors.

Many people have influenced our thoughts on experimental design over the years, and as a result some of the examples will have been borrowed in whole or in part. While it is impossible to trace and thank the originators of all these ideas, there are some people we should certainly mention. Graeme learnt a huge amount about designing experiments from Will Cresswell and Rik Smith. Kate Lessells and Andrew Read have both, over the years, forced Nick to think far harder about experiments than he ever really intended; they will both find large elements of themselves throughout the book. Anyone else that recognizes, and wishes to claim as their own, examples contained in this book is welcome to contact us. If the publishers ever let us write a second edition, we will make sure you get the proper credit.

The figures were drawn by the imaginative and efficient Liz Denton. It's been an absolute pleasure working with her. Stuart Humphries kindly provided the drawing of a bird for Figure 2.1. Thanks also to Ruedi Nager, Stuart Humphries and James Brown of the Scottish Agricultural Science Agency for cover illustrations.

Lastly we'd like to thank each other. Co-writing a book is much easier than working alone. We both found that the other person can really pick you up whenever inspiration disappears and writers' block is looming. But best of all, no matter what mistakes or shortcomings you find in this book, we'll both be able to blame it on the other guy!

Acknowledgements for the second edition

The figures for this new edition have again been produced with skill and good grace by Liz Denton. The new cover illustrations are from J. Crowe. We are extremely grateful to Lyn Knott for very careful and imaginative editing of the second edition. Thanks also to Emma Cunningham, Alex Hall and Tom Little for reading parts of the manuscript.

Contents

6 Final thoughts

Why you need to care about design

1

1.1 Why experiments need to be designed

When life sciences students see the phrase 'experimental design', it can either ease them gently into a deep sleep or cause them to run away screaming. For many, experimental design conjures up unhappy memories of mathematics or statistics lessons, and is generally thought of as something difficult that should be left to statisticians. Wrong on both counts! Designing simple but good experiments doesn't require difficult maths. Instead, experimental design is more about common sense, biological insight and careful planning. Having said that, it does require a certain type of common sense, and there are some basic rules. In this book, we hope to steer you towards thinking more effectively about designing experiments.

Designing effective experiments requires thinking about biology more than about mathematical calculations.

So why are many life scientists so averse to thinking about design? Part of the reason is probably that it is easy to think that time spent designing experiments would be better spent actually doing experiments. After all, the argument goes, we are biologists so let's concentrate on the biology and leave the statisticians to worry about the design and analysis. This attitude has given rise to a number of myths that you can hear from the greenest student or the dustiest professor.

Myth 1: ***It does not matter how you collect your data, there will always be a statistical 'fix' that will allow you to analyse them.***

It would be wonderful if this was true, but it is not. There are a large number of statistical tests out there, and this can lead to the false impression that there must be one for every situation. However, all statistical tests make assumptions about your data that must be met before the test can be meaningfully applied. Some of these assumptions are very specific to the particular test. If you cannot meet these, there may be a substitute test that assumes different characteristics of your data. But you may find that this alternative test only allows you to use your data to answer a different scientific question to the one that you originally asked. This alternative question will almost certainly be less interesting than your original question (otherwise why weren't you asking it in the first place?). It's also generally true that well designed experiments require simpler statistical methods to analyse them than less well designed experiments. Further, there are some basic assumptions that apply to all statistical tests, and you ignore these at your peril. For instance, statistical tests generally assume that your data consist of what statisticians refer to as **independent data points** (more on this in Chapter 3). If your data don't meet this criterion then there is nothing that statistics can do to help you, and your data are useless. Careful design will allow you to avoid this fate.

Independent data points come from unconnected individuals. If the measured value from one individual gives no clue as to which of the possible values the measurement of another individual will produce, then the two measurements are independent.

Careful experimental design at the outset can save a lot of sweat and tears when it comes to analysing your data.

The group of experimental subjects that we use in our experiment (called the sample) needs to be representative of the wider set of individuals in which we are interested (called the population). One key way to achieve this is to ensure that each individual selected is not linked in some way to another individual in the sample, i.e. they are independent. For example, if we were surveying human food preferences, then gathering data from five members of the same family would not produce five independent data points. We would expect that members of the same family are more likely to share food preferences than two unconnected individuals, since family members often have a long history of eating together. Similarly, gathering data from the same person on five separate occasions certainly does not provide five independent data points, since a person's preference on one occasion is likely to be a good guide to their preferences a little while later.

? **Q 1.1** If we wanted to measure the prevalences of left-handedness and religious practices among prison inmates, what population would we sample from?

? **Q 1.2** If we find that two people in our sample have been sharing a prison cell for the last 12 months, will they be independent samples?

Myth 2: ***If you collect lots of data something interesting will come out, and you'll be able to detect even very subtle effects.***

It is always reassuring to have a notebook full of data. If nothing else, it will convince your supervisor that you have been working hard. However, quantity of data is really no substitute for quality. A small quantity of carefully collected data, which can be easily analysed with powerful statistics, has a good chance of detecting interesting biological effects. In contrast, no matter how much data you have collected, if it is of poor quality, it will be unlikely to shed much light on anything. More painfully, it will probably have taken far longer and more resources to collect than a smaller sample of good data.

In science, hard work is never a substitute for clear thinking.

1.2 The costs of poor design

1.2.1 Time and money

Any experiment that is not designed in an effective fashion will at best provide limited returns on the effort and resources invested, and at worst will provide no returns at all. It is obvious that if you are unable to find a way to analyse your data, or the data that you have collected do not enable you to answer your question, then you have wasted your time and also any materials. However, even if you don't make mistakes of this magnitude, there are other ways that a poorly designed experiment might be less efficient. It is a common mistake to assume that an experiment should be as big as possible, but if you collect more data than you actually need to address your question effectively, you waste time and money. At the other extreme, if your experiment requires the use of expensive consumables, or is extremely time consuming, there is a temptation to make it as small as possible. However, if your experiment is so small that it has no chance at all of detecting the effects that you are interested in, you have saved neither time nor money, and you will probably have to repeat the experiment, this time doing it properly. Problems like these can be avoided with a bit of careful thought, and in later chapters we will discuss ways of doing so.

Similarly, it is not uncommon for people to collect as many different measurements on their samples as possible without really thinking about why they are doing so. At best this may mean that you spend a great deal of time collecting things that you have no use for, and at worst may mean that you don't collect the information that is critical for answering your question, or that you do not give sufficient time or concentration to collecting the really important information.

Don't be over-ambitious: better that you get a clear answer to one question than a guess at the answers to three questions.

Thus, while it is always tempting to jump into an experiment as quickly as possible, time spent planning and designing an experiment at the outset will save time and money (not to mention possible embarrassment) in the long run.

In science, as in life: more haste; less speed.

1.2.2 Ethical issues

If the only issue at stake in ill-conceived experiments was wasted effort and resources that would be bad enough. However, life science experiments have the additional complication that they will often involve the use of animals. Experimental procedures are likely to be stressful to animals; even keeping them in the laboratory or observing them in the wild may be enough to stress them. Thus, it is our duty to make sure that our experiments are designed as carefully as possible, so that we cause the absolute minimum of stress and suffering to any animals involved. Achieving this will often mean using as few animals as possible, but again we need to be sure that our experiment is large enough to have some chance of producing a meaningful result.

There are often many different ways that a particular experiment could be carried out. A common issue is whether we apply several treatments to the same individuals, or apply different treatments to different individuals. In the former case we could probably use fewer individuals, but they would need to be kept for longer and handled more often. Is this better or worse than using more animals but keeping them for less time and handling them less often? The pros and cons should be weighed up before the experiment is done to ensure that suffering is minimized. Issues such as this will be explored in more detail in Chapter 4.

Ethical concerns do not only apply to scientists conducting experiments on animals in a laboratory. Field experiments too can have a detrimental effect on organisms in the environment that the human experimenters are intruding on. There is no reason to expect that such an effect would even be confined to the organism under study. For example, a scientist sampling lichens from hilltops can disturb nesting birds or carry a pathogen from one site to another on their collection tools.

We cannot emphasize too strongly that, while wasting time and energy on badly designed experiments is foolish, causing more human or animal suffering or more disturbance to an ecosystem than is absolutely necessary is inexcusable.

1.3 The relationship between experimental design and statistics

It might come as a surprise that we are not going to talk about statistical analysis of data in any detail in this book. Does this mean that we think that statistical tests are not important to experimental design and that we don't need to think about them? Absolutely not! Ultimately, experimental design and statistics are intimately linked, and it is essential that you think about the statistics that you will use to analyse your data before you collect them. As we have already said, every statistical test will have slightly different assumptions about the sort of data that it requires or the sort of hypothesis that it can test, so it is essential to be sure that the data that you are collecting can be analysed by a test that will examine the hypothesis that you are interested in. The only way to be sure about this is to decide in advance how you will analyse your data when you have collected it. Thus, whilst we will not dwell on statistics in detail, we will try to highlight throughout the book the points in your design where thinking about statistics is most critical.

There are two reasons why we will not concentrate on statistics in this book. The first is simply that there are already some very good statistics books available, that you can turn to when you are at the stage of deciding what test you are planning to do (see the Bibliography for a guide to some that we have used). However, the second, more important, reason is that we believe strongly that experimental design is about far more than the statistical tests that you use. This is a point that can often be lost among the details and intricacies of statistics. Designing experiments is as much about learning to think scientifically as it is about the mechanics of the statistics that we use to analyse the data once we have them. It is about having confidence in your data, and knowing that you are measuring what you think you are measuring. It is about knowing what can be concluded from a particular type of experiment and what cannot.

Experimental design is about the biology of the system, and that is why the best people to devise biological experiments are biologists themselves.

1.4 Why good experimental design is particularly important to life scientists

Two key concepts that crop up continually when thinking about experiments (and consequently throughout this book) are **random variation** and **confounding factors**. Indeed it might be said that the two major goals of designing experiments are to minimize random variation and account for confounding factors. Both of these subjects will be covered in depth in Chapters 3 and 4 respectively. However, we mention them here in order to give a flavour of why experimental design is particularly important to life scientists.

1.4.1 Random variation

Random variation is the differences between measured values of the same variable taken from different experimental subjects.

Random variation is also called between-individual variation, inter-individual variation, within-treatment variation or noise. This simply quantifies the extent to which individuals in our sample (which could be animals, plants, human subjects, study plots or tissue samples, to give but a few examples) differ from each other. We discuss the consequences of this fully in Chapter 3. For example, all 10-year-old boys are not the same height. If our study aims to study national differences in the height of 10-year-old boys, we should expect that all boys in our study will not have the same height, and that this variation in height will be driven by a large number of factors (e.g. diet, socio-economic group) as well as the possible effect of the factor we are interested in (nationality).

Q 1.3 Humans have tremendous variation in the patterning of grooves on our fingers, allowing us to be individually identified by our fingerprints. Why do you think there is such variation?

Random variation is everywhere in biology. All the sheep in a flock are not exactly the same. Hence, if we want to describe the characteristic weight of a sheep from a particular flock, we generally cannot simply weigh one of the sheep and argue that its weight is also valid for any and all sheep in the flock. It would be better to measure the weights of a representative sample of the flock, allowing us to describe both the average weight of sheep in this flock and the extent of variation around that average. Things can be different in other branches of science: a physicist need only calculate the mass of one electron, because (unlike sheep) all electrons are the same weight. Physicists often need not concern themselves with variation, but for life scientists it is an ever-present concern, and we must take it into account when doing experiments.

Good experiments minimize random variation, so that any variation due to the factors of interest can be detected more easily.

1.4.2 Confounding factors

If we want to study the effect of variable *A* on variable *B*, but variable *C* also affects *B*, then *C* is a **confounding factor**.

Often we want to understand the effect of one factor (let's call it variable *A*) on another factor (*B*). However, our ability to do this can be undermined if *B* is also influenced by another factor (*C*). In such circumstances, *C* is called a **confounding factor** (sometimes referred to as a confounding variable or third variable). For example, if we observe juvenile salmon in a stream to see if there is an effect of the number of hours of sunlight in the day (variable A) on rates of foraging (variable B), then water temperature (variable C) may be a confounding factor. We might expect increased sunlight to increase foraging activity, but we might also expect increased water temperature to have the same effect. Disentangling the influences of temperature and sunlight is likely to be particularly problematic as we might expect sunlight and water temperature to be closely linked, so it will be challenging to understand the effect of one in isolation from the other. Hence confounding factors pose a challenge, but not an insurmountable one, as we'll discuss fully in Chapter 4.

Another advantage that the physicist has is that he or she can often deal with very simple systems. For example, they can isolate the electron that they want to measure in a vacuum chamber containing no other particles that can interact with their particle of interest. The life scientist generally studies complex systems, with many interacting factors. The weight of an individual sheep can change quite considerably over the course of the day through digestion and ingestion, or through being rained on. Hence, something as simple as taking the weight of animals is not as simple as it first appears. Imagine that today you measure the average weight of sheep in one flock and tomorrow you do the same for another flock. Let's further imagine that you find that the average weight is higher for the second flock. Are the animals in the second flock really intrinsically heavier? Because of the way that you designed your experiment it is hard to know. The problem is that you have introduced a confounding factor: time of measurement. The sheep in the two flocks differ in an unintended way; they were measured on different days. If it rained overnight then there might be no real difference between sheep in the two flocks; it might just be that all sheep (in both flocks) are heavier on the second day because they are wetter. We need to find ways to perform experiments that avoid or account for confounding factors, so that we can understand the effects of the factors that we are interested in. Techniques for doing this will be tackled in Chapter 4. However, before we can do this, we have to be clear about the scientific question that we want a given experiment to address. Designing the right experiment to answer a specific question is what the next chapter is all about.

Q 1.4 If we are interested in comparing eyesight between smokers and non-smokers, what other factors could contribute to variation between people in the quality of their eyesight? Are any of the factors that you've chosen likely to influence someone's propensity to smoke?

Q 1.5 Faced with two flocks of sheep 25 km apart, how might you go about measuring sample masses in such a way as to reduce or remove the effect of time as a confounding factor?

Confounding factors make it difficult for us to interpret our results, but their effect can be eliminated or controlled by good design.

Summary

- You cannot be a good life scientist without understanding the basics of experimental design.

- **The basics of experimental design amount to a small number of simple rules; you do not have to get involved in complex mathematics in order to design simple but good experiments.**
- **If you design poor experiments, then you will pay in time and resources wasted.**
- **Time and resource concerns are trivial compared to the imperative of designing good experiments so as to reduce (or hopefully eliminate) costs to your experiment in terms of suffering to animals or humans or disturbance to an ecosystem.**

Starting with a well-defined hypothesis

2

Your aim in conducting an experiment is to test one or more specific scientific hypotheses.

- There is no way that you can hope to design a good experiment unless you first have a well-defined hypothesis (see section 2.1).
- One aspect to good design is to produce the strongest test of the hypothesis (section 2.2).
- You should be aiming to produce results that convince even sceptics (section 2.3).
- A pilot study will allow you to focus your aims and perfect your data collection techniques (section 2.4).
- A key decision in your design is likely to be whether to perform an experimental manipulation or simply record natural variation. Both techniques have their advantages and disadvantages, and a combination of the two is often effective (section 2.5).
- You may then have to address whether to study your subjects in their natural environment or in the more controlled conditions of the laboratory (section 2.6) or whether to carry out experiments *in vitro* or *in vivo* (section 2.7).
- All these decisions involve an element of compromise, and while there are good and bad experiments (and all the shades in between), there are no perfect ones (section 2.8).

2.1 Why your experiment should be focused: questions, hypotheses and predictions

A **hypothesis** is a clear statement articulating a plausible candidate explanation for observations. It should be constructed in such a way as to allow gathering of data that can be used to either refute or support this candidate explanation.

A **hypothesis** is a clearly stated postulated description of how an experimental system works.

There can be several hypotheses for the same observation. For example, take the observation that patients in the beds nearest to the entrance to a hospital ward respond more effectively to treatment than those at the other end of the ward. Candidate hypotheses could include the following:

Q 2.1 Suggest some hypotheses that could explain the observation that people drive faster on the journey to work than on the way home.

1. Patients are not distributed randomly to beds, with the ward sister having a tendency to place more severe cases at the quieter end of the room away from the entrance.
2. Patients nearer the door are seen first on the doctor's round and end up getting more of the doctor's time and attention.
3. Patients respond positively to increased levels of activity and human contact associated with being nearer the ward entrance.

Don't get yourself into a position where you think something like,

> 'Chimps are really interesting animals, so I'll go down to the zoo, video the chimps and there are bound to be lots of interesting data in 100 hours of footage.'

We agree that chimpanzees can be very interesting to watch, so by all means watch them for a while. However, use what you see in this **pilot study** to generate clear questions. Once you have your question in mind, you should try to form hypotheses that might answer the question. You then need to make predictions about things you would expect to observe if your hypothesis is true and/or things that you would not expect to see if the hypothesis is true. You then need to decide what data you need to collect to either confirm or refute your predictions. Only then can you design your experiment and collect your data. This approach of having specific questions, hypotheses and predictions is much more likely to produce firm conclusions than data collected with no specific plan for its use. This should be no surprise; if you don't know how you are going to use data, the chance of you collecting a suitable amount of the right sort of data is too slim to take a chance.

A **pilot study** is an exploration of the study system that is conducted before the main body of data collection in order to refine research aims and data collection techniques. A good pilot study will maximize the benefits of your main data-collection phase and help you avoid pitfalls.

Given the importance of a focused research question, you might be thinking that spending even a few minutes making unfocused observations of the chimps is simply wasted time that could be better spent collecting data to answer scientific questions. We disagree. You should be striving to address the most interesting scientific questions you can, not as many questions as you can. Darwin, Einstein, Watson and Crick are not famous because they solved more problems than anyone else, but because they solved important problems. So take the time to think about what the interesting questions are. We are not promising that you'll become as famous as Einstein if you follow this advice, but the chances will increase.

As a general rule your experiment should be designed to test at least one clear hypothesis about your research system. A period of unfocused

observation can help you identify interesting research questions, but a study based entirely on unfocused observation will rarely answer such questions.

2.1.1 An example of moving from a question to hypotheses, and then to an experimental design

The last section might have seemed a little abstract, so let us try an example. Imagine that you spend some time at the zoo doing a pilot study watching the chimpanzees. During this time you notice that the chimps show strong daily variation in activity patterns. This leads naturally to the research question:

> Why does chimp activity vary during the day?

Our observations have allowed us to come up with a clear question, and with a clear question we have at least a fighting chance of coming up with a clear answer. The next step is to come up with hypotheses that could potentially explain our observation. One such hypothesis might be:

> Chimp activity pattern is affected by feeding regime.

Now, how would we test this hypothesis? The key is to come up with a number of predictions of observations that we would expect to make if our hypothesis is correct. So a prediction has to follow logically from the hypothesis *and* has to be something that we can test. In the case of this hypothesis, we might make the prediction:

> The fraction of time that a chimp spends moving around will be higher in the hour around feeding time than at other times of the day.

By making a clear prediction like this, it becomes obvious what data we need to collect to test our prediction. If our prediction is supported by those data, then this provides some support for the hypothesis (assuming that we have come up with a sensible prediction that tests our hypothesis). In statistics texts, you'll find hypotheses linked as pairs, with each pair being composed of a null hypothesis and an alternative hypothesis: Statistics Box 2.1 explains this concept further.

In order for your hypothesis to be useful, it has to generate *testable* predictions.

2.1.2 An example of multiple hypotheses

There is nothing to say that you are limited to a single hypothesis in any particular study. Let's consider a second example. You are walking along

STATISTICS BOX 2.1 Null and alternative hypotheses

Statisticians talk about two different kinds of hypothesis, a null hypothesis and an alternative hypothesis. We mention these because you will come across them in the statistical literature. The null hypothesis can be thought of as the hypothesis that nothing is going on. In the example in section 2.1.1, a possible null hypothesis would be:

Chimp activity is not affected by feeding regime.

In effect, what the null hypothesis is saying is that any apparent relationship between chimp activity and feeding regime is just due to chance. Such a null hypothesis would lead to the prediction:

There is no difference in the fraction of time that the chimps spend moving in the hour around feeding compared to the rest of the day.

The hypothesis that we originally gave is an example of an alternative hypothesis, i.e. a hypothesis in which the pattern that we see is due to something biologically interesting, rather than just being due to chance. The reason that we need a null hypothesis is right at the heart of the way in which statistics work, but the details needn't bother us too much here. However, it boils down to the fact that statistical tests work by testing the null hypothesis. If our observations allow us to reject the null hypothesis (because the data we collect disagree with a prediction based on the null hypothesis), then this gives us support for our alternative hypothesis. So if we can reject the null hypothesis that chimp activity is not affected by feeding regime, logically we should accept that chimp activity *is* affected by feeding regime.

This may seem rather a twisted way to approach problems, but there is good reason behind this approach. Philosophically, science works conservatively. We assume that nothing interesting is happening unless we have definite evidence to the contrary. This stops us getting carried away with wild speculations, and building castles in the air. This philosophy leads to concentration on the null hypothesis rather than the alternative hypothesis. However, while this philosophy is the bedrock of statistical testing, it need not detain us further here. Just remember that for every hypothesis you make that suggests that something interesting is happening, there will be a corresponding null hypothesis (that nothing interesting is happening), and that by testing this null hypothesis you can gain information about the hypothesis that you are interested in.

a rocky shore and you notice that the whelks on the rocks often seem to occur in groups. You come up with the sensible question:

Why do whelks group?

Your first thought is that it might be to do with seeking shelter from the mechanical stresses imposed by breaking waves, so you generate a hypothesis:

Whelks group for shelter from wave action.

Then, after a little more thought, you wonder if maybe grouping is something to do with clumping in areas of high food, and this leads you to a second hypothesis:

Whelks group for feeding.

As with the previous example, we now need to come up with predictions that we can test. Ideally, we would like predictions that will allow us to discriminate between these two hypotheses. If the first hypothesis is true, we might predict:

Whelks are more likely to be found in groups in areas sheltered from wave action.

whereas if the second is true, our prediction would be:

Whelks are more likely to be found in groups in areas of higher food density.

With these predictions in hand, we could then design a study to test both of them. Of course, even if we find one of our predictions to be supported, for example that whelks are more likely to be found in groups in areas sheltered from wave action, this does not necessarily mean that the hypothesis that led to the prediction ('whelks group for shelter from wave action') is correct. Someone might come up with another hypothesis that would also lead to this prediction. For example, the hypothesis:

Whelks are more vulnerable to predators in sheltered areas, but grouping provides protection from predators

could also lead to the prediction:

Whelks are more likely to be found in groups in areas sheltered from wave action.

Your task will then be to come up with another prediction that can discriminate the new hypothesis from the original one. Now that is not to say that both hypotheses cannot be true. Maybe part of the reason that we find aggregations of whelks in sheltered areas is to avoid wave action, and part of it is to avoid predators in the sheltered areas. Indeed, it is common in the life sciences to find that several hypotheses can be combined to fully explain observations. The key thing is that if only one hypothesis is correct, we should be able to determine that from our study. To clarify what we mean lets call these two hypotheses *predation* and *shelter*. We then have four possibilities for how the system actually works.

Possibility 1 Neither hypothesis is true and the observed patterns are due to something else entirely.

Possibility 2 *Predation* is true and *shelter* is false.

Possibility 3 *Shelter* is true and *predation* is false.

Possibility 4 Both *predation* and *shelter* are true.

A study that allowed us to discriminate possibility 1 from 2, 3, and 4, but did not allow discrimination between 2, 3, and 4, would be useful, but not ideal. The ideal study would allow us not only to discriminate possibility 1 from the others, but also to discriminate between these other possibilities. That is, at the end of the study, we could identify which of the four possible combinations of the two hypotheses is actually correct. Such a study might include a series of experimental treatments where either predators or wave action or both were removed and the grouping behaviour of the whelks observed, or it might involve making observations on beaches with different combinations of predation and wave action. No matter how the study is organized, the important thing is that the best study will be the one that allows us to tease apart the influence of the different hypothesized influences on grouping behaviour.

Q 2.2 A report claims that a questionnaire survey demonstrates that those who reported regularly playing computer games displayed attitudes consistent with an increased propensity to perform violent acts. Suggest some hypotheses that could explain this association and how predictions from these hypotheses might be tested.

Generating sensible predictions is one of the 'arts' of experimental design. Good predictions will follow logically from the hypothesis we wish to test, and hopefully not from other rival hypotheses. Good predictions will also lead to obvious experiments that allow the prediction to be tested. Going through the procedure of

question→hypothesis→prediction

is an extremely useful thing to do, in that it makes the logic behind our study explicit. It allows us to think very clearly about what data we need to test our predictions, and so evaluate our hypothesis. It also makes the logic behind the experiment very clear to other people. We would recommend strongly that you get into the habit of thinking about your scientific studies this way.

2.2 Producing the strongest evidence with which to challenge a hypothesis

Try to find the strongest test of a scientific hypothesis that you can. Consider the hypothesis:

> Students enjoy the course in human anatomy more than the course in experimental design.

You might argue that one way to test this hypothesis would be to look at the exam results in the two courses: if the students got higher grades in one particular course compared to another, then they probably enjoyed it more, and so were more motivated to study for the exam. Thus our prediction, based on our hypothesis, is:

Students will get higher marks in the human anatomy exam than in the experimental design exam.

We would describe this as a weak test of the hypothesis. Imagine that students do score significantly higher in human anatomy. This is *consistent* with the hypothesis. However, it could also be explained by lots of other effects: it may be that the human anatomy exam was just easier, or that it fell at the start of the set of exams when students were fresh, whereas experimental design came at the end when they were exhausted. This problem of alternative explanations is explored further in section 2.5.3.

Data that are consistent with a hypothesis are less powerful if those data are also consistent with other plausible hypotheses.

2.2.1 Indirect measures

An **indirect measure** is a measure taken on a variable that we are not primarily interested in but which can be used as an indicator of the state of another variable that is difficult or impossible to measure.

The problem with using exam scores to infer something about student's enjoyment levels is that you are using an **indirect measure**. This involves measuring one variable as a surrogate for another (generally more difficult to measure) variable. For example, if we are interested in comparing the amount of milk provided to seal pups by mothers of different ages, then we might consider that the amount of time that pups spend suckling is a reasonable surrogate for the quantity of milk transferred. Although this makes data collection much easier, you must be sure that we are using a good surrogate. If rate of milk delivery during suckling varies with age of the mother, then time spent suckling could be a poor measure of actual quantity of milk transferred.

Indirect measures generally produce unclear results, as in the case of students' exam performance above. If we can, we should always try to measure the thing we are interested in directly. In the course enjoyment study, we should interview the students (separately; see section 3.3 on independence of data points) and ask them which course they prefer. However, if the questioner were the human anatomy lecturer, then we should worry about how honest the answers would be. It would be better to have the questions asked by someone who the students perceive to be disinterested in the answer.

Q 2.3 In the original study referred to in Q 2.2, the investigators were interested in people's propensity to commit violent acts. However, they used an indirect measure of this by asking people to answer a carefully designed questionnaire about their attitude to certain situations, such that these answers could be used to infer propensity to violence. Why do you think this approach was taken rather than using a direct measure of propensity to violence?

Don't use indirect measures unless you have to.

2.2.2 Considering all possible outcomes of an experiment

One moral of the last couple of pages is that, before doing an experiment, you should always ask yourself, for *every possible outcome*, how such a set of results could be interpreted in terms of the hypotheses being tested. Try to avoid experiments that can produce potential outcomes that you cannot interpret. It doesn't make you sound very impressive as a scientist if you have to report that:

> I collected data this way, the data look like this, and I have no idea what I should conclude about my study system from this.

It can sometimes be difficult to think of every possible outcome of an experiment beforehand, and again, pilot studies can be an effective way of uncovering these.

Another aspect of this warning is to be wary of doing experiments where a useful outcome of the work hangs on getting a specific result. There is nothing wrong with thinking, 'Lets do this experiment, because if our hypothesis is supported by the experiment, then we'll really cause a sensation'. However, make sure that your results are still interesting and useful if the outcome is not the one that you hoped for.

Imagine that your experiment is looking to explore the effect of one factor on another. What if you find no effect at all? You must ask yourself whether someone else would still judge your experiment to have been scientifically valid and a useful way to spend time and resources even if you get this negative result. For example, imagine that you performed an effective investigation to test the hypothesis:

> Cannabis use has an effect on driving ability.

If you find no relationship between the two, then that would be an interesting result. It is interesting because it conflicts with our current understanding of the effects of this substance. It would be relevant both to our fundamental understanding of the effect of the drug on the brain and to the scientific underpinning of motoring law related to drug use. If instead you find an effect, and can quantify the effect, then this should be an interesting result too, again with implications for our understanding of brain function and for legislation.

Now imagine that you performed an effective investigation to test the hypothesis:

> Preference for butter or margarine is linked to driving ability.

If you find strong evidence supporting this hypothesis, then this would be surprising and interesting, as it conflicts with our current understanding of the factors affecting driving ability. However, if you find no relationship between whether people prefer butter or margarine and their driving ability, then this is a less interesting result. People could simply say that you were never going to find an effect between two such entirely unrelated factors and that you should have spent your time on more productive lines of enquiry.

We are not saying, 'don't ever do an experiment where one potential result of the experiment is uninteresting'. If you had demonstrated that a preference for butter is associated with better driving, then that could be very interesting indeed. You might consider that the fame and fortune associated with discovering such results are worth the risk of generating lots of dull results. It's your decision; our job is to point out to you that there are risks.

Think about how you would interpret every possible outcome of your experiment, before you decide to do the experiment.

2.3 Satisfying sceptics: the Devil's advocate

One theme running through this book is that your experimental design and data collection should satisfy someone that we describe as 'the Devil's advocate'. By this, we mean that you must make sure that your experiment is designed so that the conclusions that you draw are as strong as possible. Let's imagine that you are in a situation where a group of observations could be explained by either of two contradictory mechanisms (A or B), and your aim is to try and illuminate which is in fact the real mechanism underlying the results. Don't be satisfied with an experiment that will let you conclude:

> The results of this experiment strongly suggest that mechanism A is operating. While it is true that the results are equally as compatible with mechanism B, we feel that A is much more likely.

You have to expect the person refereeing your manuscript or marking your project report to be a 'Devil's advocate', who constantly says, 'No, I'm not prepared to give you the benefit of the doubt, prove it'. For the case above, their conclusion is:

> It is a matter of opinion where mechanism A or B is more likely, hence I can conclude nothing from your experiment.

It would be much better to design an experiment that would allow you to conclude something stronger, such as:

> These results are entirely consistent with the hypothesis that mechanism A is in operation. However, they are totally inconsistent with mechanism B. Therefore we conclude that mechanism A is the one acting.

The Devil's advocate is wilfully looking to shoot holes in your argument, and you must restrict their ammunition. The Devil's advocate cannot argue that mechanism B is operating if you have proved that it is not.

You should think of the Devil's advocate as a highly intelligent but sceptical person. If there is a weakness in your argument, then they will find it. However, they can be made to believe you, but only when they have no reasonable alternative. They will not give you the benefit of any reasonable doubt.

2.4 The importance of a pilot study and preliminary data

It seems inevitable that as soon as you begin a scientific study you will find yourself under time pressure. Undergraduate projects are generally only brief, a PhD never seems to last long enough and the end of a post-doctoral grant begins looming on the day that it starts. Such pressures can lead to a temptation to jump straight in and do some experiments quickly. Having reams of data can be very comforting. However, the old adage of 'more haste; less speed' applies to biological studies as much as anywhere, and an important step in experimental biology that is often missed is spending a bit of time at the beginning of a study collecting some pilot observations. A pilot study can mean anything from going to the study site and watching for a few hours or days, to trying a run of your experiment on a limited scale. Exactly what it entails will depend on the particular situation. Leaping straight into a full-scale study without carrying out such pilot work will often mean that you get less from the study than you might have done if the pilot study had suggested some fine-tuning. Not infrequently, your 3-month long full-scale study will be totally useless because of a pitfall that a 3-day pilot study would have alerted you to. In section 2.1 we suggested that a pilot study can be invaluable in helping you to develop interesting, focused questions. Over the next few pages we will outline several other advantages of pilot studies.

2.4.1 Making sure that you are asking a sensible question

One aim of a pilot study is to allow you to become well acquainted with the system that you will be working on. During this phase of a project, you will be trying to gain information that will help you to better design your main experiments. But what should you be aiming to do in this time?

Probably the most important goal of a pilot study is to validate the biological question that you are hoping to address. You may have read about stone-stealing behaviour in chinstrap penguins, a phenomenon where penguins steal stones from each others' nests and incorporate them into their own nest. Let's say that you decide that this would make an interesting subject for your investigation. Fortunately, there is a chinstrap colony at the local zoo that you will be able to work on. So you form hypotheses about aspects of stone stealing, and design an elaborate experimental programme to evaluate these hypotheses. This involves placing all sorts of manipulations on penguins' nests, to add and remove stones of different sizes and colours. Setting up the experiment is hard work, and the penguins suffer considerable disturbance. However, eventually it is done and you sit back in your deck-chair to begin collecting the data. After a day, you have seen no stone-stealing behaviour, but you are not bothered: stone stealing is a well-documented phenomenon, and there are plenty of papers written on the subject. Unfortunately, your particular penguins haven't read these papers, and after a month of observing and not seeing a single case of stone stealing you give up.

If only you had taken the time to go and watch the penguins for a couple of days before putting in all your hard work. You would have realized that stone stealing was a rare occurrence in this colony, and that you might have to rethink your project. You might also have seen some other interesting behaviour that could have formed the basis of a valuable alternative study. Alternatively, you might have observed two very different types of stealing behaviour, allowing you to plan your project in such a way that you can look at these two different types, making for an even better project.

It is worth mentioning at this point the dangers of relying on others to give you this information, no matter how august they are, unless you absolutely have to. Your supervisor might tell you that stone stealing will occur in your study colony. Maybe they are basing this on having read the same papers that you read, or on a discussion with a zoo keeper, who, on the basis of having seen a case 2 years ago, proudly asserts that stone stealing goes on at their colony. At the very least you should enquire when the behaviour was seen and how often. In the end, it is your project; you will be the one putting in the hard work and wasting time if it goes wrong, so you should make the preliminary observations that convince you that your main study is possible. Some things you have to take on trust, but don't take things on trust if you can easily check them for yourself.

It may be that you consider our unfortunate penguin watcher to be a straw man, and that there is no way that you would be so foolish. We can only say that we have encountered a large number of analogous abortive experiments and have been the guilty party ourselves on enough occasions to make us squirm.

Pilot studies save the potentially considerable embarrassment, wasted expense and perhaps animal suffering of such incidents.

Don't take things on trust that you can easily check for yourself.

2.4.2 Making sure that your techniques work

A second crucial role for a pilot study is that it gives you a chance to practise and validate the techniques that you will use in the full study. This might involve something as simple as making sure that you will be able to make all the observations on the penguins that you need from the comfort of your deck-chair, or that you won't disturb the animals too much. You might also decide to check that you can use the shorthand code that you have devised to allow you to follow the behaviour of the birds, and that you have a code for all the behaviours you need. Or you might need to check that you can take all the measurements you need in the time that you have available. Of course, this doesn't just apply to field projects. You may have planned an experiment that requires you to count bacteria with a microscope, but how long can you spend staring down a microscope before your accuracy deteriorates? The only way to know is to try it, and this will ultimately affect the maximum size of the experiment you can plan for and carry out.

There is a lot to be said for practising all the techniques and methods that you will use in your experiment before you use them in your main experiment. If your experiment will involve dissecting mosquitoes, you want to make sure that you can do this efficiently, so as to be sure that you will not destroy valuable data points and cause unnecessary suffering.

Trial runs of different types of assays can also allow you to fine-tune them so that they will give you useful results. You might be planning a choice test for female butterflies in which you will give a female a choice of several host plants and see which she lays her eggs on. If you leave her for too long to lay eggs, she may lay eggs on all the hosts, and subtle preferences will be hard to detect. Conversely, if you leave her for too little time, she may not have sufficient time to properly explore all the hosts, so may lay eggs at random. By carrying out pilot experiments with females left for different lengths of time, you can work out what time interval to use in the main experiment.

Pilot studies also provide an opportunity for techniques to be standardized in studies involving several observers. Perhaps you need to check that what one person regards as an aggressive behaviour is the same for all observers, or that all are measuring things in the same way or from the same positions. See Chapter 5 for more on this. Another advantage of a pilot study is that it can provide a small-scale data set, of the same form as your main planned experiment will produce. This gives you a chance to test your planned statistical analysis; see Statistics Box 2.2 for more on this.

STATISTICS BOX 2.2 **Obtaining basic data to fine-tune design and statistics**

An advantage of carrying out a pilot study in the form of a small-scale version of the actual experiment that you have planned is that, if your techniques work well, you will have some preliminary experimental data of the sort that you will collect in the full study. This can be extremely valuable in two ways.

Firstly, such data can provide the necessary information to see whether the planned size of the experiment is likely to be appropriate. This will be discussed in more detail when we describe statistical power in Chapter 3.

Secondly, these data can give you something to try out your planned statistical tests on. Whenever you carry out a study, you should have planned how you will analyse the data that you collect, and the best way to decide whether your analysis will work is to try it on a subset of data from a pilot experiment. Every statistical test requires data of a specific type that must meet a number of assumptions, and there is nothing worse than finding at the end of a study that you have collected your data in a way that makes it impossible to analyse (or much more difficult than was necessary). Such problems will immediately become apparent if you try to analyse your pilot data, allowing you to modify the way you do your study before it is too late. This pilot analysis also has yet another advantage. Like most things, manipulating data and performing statistical tests gets easier with practice. Once you have learnt how to do the appropriate specific analysis on your pilot data, it will then be easy to do the same analysis on the full data set. It will almost inevitably be the case that your data collection takes slightly longer to do than you had planned, leaving you less time at the end of the study to analyse the data and write the results up. If this is the point when you first worry about statistics, this is likely to put you under even more time pressure. If, on the other hand, you have already dealt with all the details of the statistics early in the study, the analysis will take much less effort at the end, allowing you to spend more time writing and thinking about what your results mean. Indeed, even in those situations where you can't actually obtain pilot data yourself, we would encourage you to either borrow some data from someone else, or make up a dummy data set of the sort that you expect to collect, in order to try your analysis out. With statistics, as with all the other techniques that you use, the best way to ensure that they will work when they need to is to try them first when the pressure is off.

Q 2.4 The research task you've been assigned is to go to a certain pedestrian walkway over a motorway and count the number of cars passing underneath you, so as to test the hypothesis that more cars travel from east to west than from west to east on this stretch of motorway between 8 am and 9 am on weekdays. What aspects of this study would you seek to evaluate in a pilot study before you begin?

In short, many of the problems that might occur in the main experiment can be ironed out by careful thought beforehand. However, nothing really brings out all the possible flaws and problems with an experiment like trying out the techniques under the conditions that you will use them. We encourage you to do this whenever possible.

Practise all your data collection techniques in a realistic setting before you need to use them for real.

2.5 Experimental manipulation versus natural variation

Once you have chosen your broad research question, you then face the problem of how you are going to test your specific hypotheses. This is the crux of experimental design. One of the major decisions you will have to make is whether your study will be **manipulative** or **correlational**. A manipulative study, as the name suggests, is where the investigator actually does something to the study system and then measures the effects these manipulations have on the things that they are interested in. In contrast, a correlational study makes use of naturally occurring variation rather than artificially creating variation to look for the effect of one factor on another. A correlational study may also be called a mensurative experiment or an observational study.

In a **manipulative** study, the experimenter changes something about the experimental system and then studies the effects of this change, whereas **correlational** (**observational**) studies do not alter the experimental system.

2.5.1 An example of a hypothesis that could be tackled by either manipulation or correlation

Imagine that we had a hypothesis:

> Long tail streamers seen in many species of bird have evolved to make males more attractive to females.

From this we could make the straightforward prediction that males with long tails should get more matings than males with short tails, but how would we test this?

Correlation approach

One way to test the prediction would be to find ourselves a study site and catch a number of males. For each male that we catch, we measure the length of its tail streamers, add leg bands to the bird to allow individual recognition and then let it go again. We could then watch the birds for the rest of the season and see how many females mate with each male. If, after doing the appropriate statistical analysis, we found that the males with longer tails obtained more matings, this would support our hypothesis. A study like this is a correlational study; we have not actually manipulated anything ourselves, but made use of the naturally occurring variation in tail length to look for a relationship between tail length and number of mates.

Manipulative approach

An alternative approach would be to actually manipulate the length of the tail. So again, we would catch the male birds, but instead of just measuring them and letting them go, we would manipulate their tail length. For example, we might assign the birds into three groups. Birds in the first group have the ends of their tail streamers cut off and stuck back on again; these would form the control group (see Chapter 4 for more on controls). Birds in group 2 would have the ends of their tail streamers cut off, reducing their length. Birds in group 3 would have their tails increased in length by first cutting off the ends of their streamers and then sticking them back on, along with the bits of streamers that were removed from the birds in group 2. We then have three groups, all of which have experienced similar procedures, but one resulted in no change in streamer length, the next involves shortening of the streamers and the last involves elongating them (see Figure 2.1). We would then add the leg bands, let the birds go and monitor the numbers of matings. If the birds in the group with the extended tails got more matings than the birds in the other groups, this would support our hypothesis. If the birds that had had their streamers reduced also got fewer matings than the controls, this would also add support for our hypothesis.

In this example, both the manipulative and correlational techniques seem effective. Below we consider the factors that must be considered in order to decide which would be most effective.

Both manipulative and correlational studies can be effective, and the best approach will depend on the specifics of your situation.

2.5.2 Arguments for doing a correlational study

Correlational studies have several advantages. They are often easier to carry out than manipulative studies. Even from the number of words that we've just used to describe these two different ways of testing our hypothesis about tail streamers, it is clear that the correlational study will often involve less work. This may not just be an advantage through saved time and effort, but it may also mean that you have to handle or confine organisms for much less time, if at all. If we are dealing with organisms that are likely to be stressed or damaged by handling, or samples that can be contaminated, this is obviously a good thing.

A major worry of doing manipulative studies is that the manipulation that you have carried out will have unintended effects. Biological organisms are integrated units and any change in one part may have profound effects on other functions. If we cut off part of a male bird's tail, this might affect his ability to fly as well as his attractiveness, and any results we see as a consequence might be due to changes in his flight ability, and not due to his tail length directly. Even in fields where modern techniques of genetic engineering allow single genes to be modified or knocked out, it is impossible to be sure that our manipulation doesn't alter things other than

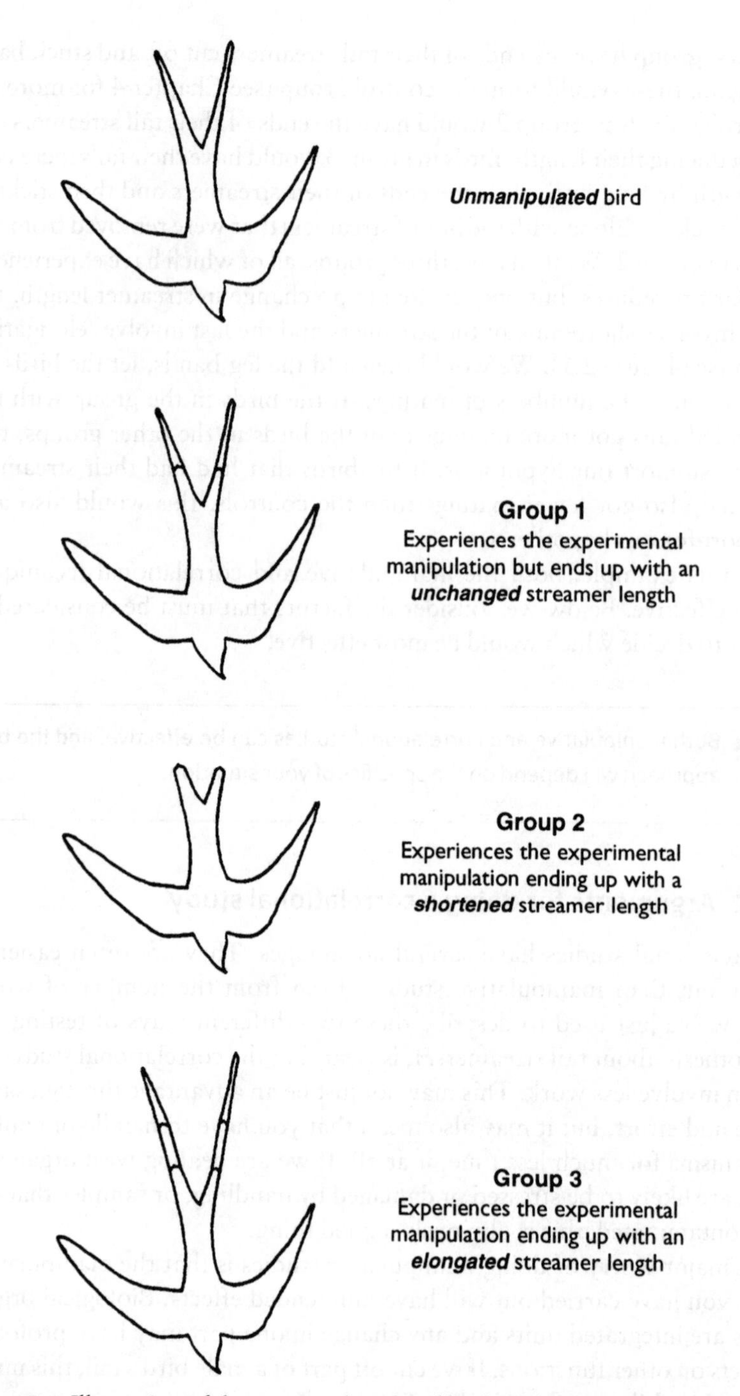

Figure 2.1 Illustration of the streamer lengths of birds in the three groups, for the experiment testing for a relationship between streamer length and mating success.

the characters we are interested in. If we design our experiments carefully, with adequate and appropriate controls, we should be able to detect such spurious effects if they occur, but we should always be cautious. Such a problem will not arise in a correlational study.

A final benefit of correlational studies is that we can be sure that we are dealing with biologically relevant variation, because we haven't altered it. Suppose that we extend or reduce tail streamers by 20 cm, but in nature tail streamers never vary by more than 2 cm across the whole population. Since we have modified birds such that they have traits that are far outside the naturally occurring range, it is doubtful that our experiment can tell us anything biologically relevant about the system. Again, this need not be a fatal problem of manipulative studies. As long as manipulations are carefully planned, and based on adequate biological knowledge of and pilot data from the system, this problem can be avoided. That said, we urge you to think carefully about the biological relevance of any manipulation that you plan before you carry it out.

Correlational studies are generally easier to perform, and have less chance than manipulative studies of going badly wrong.

2.5.3 Arguments for doing a manipulative study

Given all the advantages listed in the preceding section, you might now be wondering why anyone ever bothers to do a manipulative study at all. Despite the advantages offered by correlational studies, they suffer from two problems: **third variables** and **reverse causation**. As problems go, these can be big ones.

A **third variable** is a variable (*C*) that separately affects the two variables that we are interested in (*A* and *B*), and can cause us to mistakenly conclude that there is a direct link between *A* and *B* when there is no such link.

Third variables

Let's look at third variables first. We can sometimes mistakenly deduce a link between factor A and factor B when there is no direct link between them. This can occur if another factor C independently affects both A and B. C is the third variable that can cause us to see a relationship between A and B despite there being no mechanism providing a direct link between them. You should recognize that third variables are an example of what we referred to in Chapter 1 as confounding variables.

As an example, imagine that we survey patients in a General Practitioner's waiting room and find that those that travelled to the GP's by bus had more severe symptoms than those who travelled by car. We would be rash to conclude from this that travelling by bus is bad for your health. The cause of the correlation is likely to reside in a third variable: for example, socio-economic group. It may be that there is no direct link between bus travel and poor health, but those from lower socio-economic groups are likely to be in poorer health and more likely to use buses than the more affluent.

Let's go back to the correlational study above where we found a relationship between tail streamer length and the number of matings that a male obtained. Does this show that the length of a male's tail affects the number of matings that he gets? What if females actually base their choice on the territory quality of a male, not on his tail length at all, but males on good territories are able to grow longer tails (maybe they can get more food on the better territories). We think that we have observed a causal relationship between the number of matings that a male gets and his tail length, but what actually occurs are separate relationships between the quality of territory and both the number of matings that the male gets and his tail length. Territory quality is a third variable that affects both male tail length and the number of mates that the male gets, and because it affects both, it leads us to see a relationship between the number of matings that a male gets and the length of his tail, even though tail length does not affect females' choices at all. In almost any correlational study there will be a large number of third variables that we have not measured and that could be producing the relationship that we observe. Its important that you understand the concept of third variables clearly, so let's look at some further examples.

Ronald Fisher, a famous (and eccentric) evolutionary biologist and statistician, is reputed to have used third variables to discount the observed relationship between smoking and disease. Smoking, he argued, does not cause diseases; it is just that the types of people that smoke also happen to be the types of people that get diseases for other reasons. Maybe these are diseases caused by stress, and stressed people are also more likely to smoke.

More women graduates never marry than would be expected for the population as a whole. This does not mean that going to university makes a woman less likely to marry as such (or that no-one would want to marry the sort of male you meet at university). It perhaps means that personality traits or social circumstances associated with a slightly increased chance of going to university are also associated with a slightly decreased chance of marrying.

There is a correlation between the amount of ice-cream sold in England in a given week and the number of drowning accidents. This is unlikely to be a direct effect: that for example heavy consumption of ice-cream leads to muscle cramp or loss of consciousness. More likely a third variable is at work: when the weather is warm then more ice-creams are sold, but also (and separately) inexperienced sailors hire boats, and people who are not experienced swimmers dive into rivers and lakes.

In short, third variables mean that we can never be sure that the pattern that we have observed in an unmanipulated system is really due to the factors that we have measured and not due to correlation with some other unmeasured variable. The only way to be certain of removing problems with third variables is to carry out experimental manipulations.

Q 2.5 In a survey of road-killed badgers, it was found that those with higher levels of intestinal parasites had lower body weights: can we safely conclude that intestinal parasites cause weight loss, or is there a plausible third-variable effect that could explain this observation?

Reverse causation

The second problem of correlational studies is **reverse causation**. This can occur when we see a relationship between factors A and B. It means mistakenly assuming that factor A influences factor B, when in fact it is change in B that drives change in A.

Reverse causation is mistakenly concluding that variable *A* influences variable *B* when actually it is *B* that influences *A*.

For example, imagine a survey shows who those who consider themselves to regularly use recreational drugs also consider themselves to have financial worries. It might be tempting to conclude that a drug habit is likely to cause financial problems. The reverse causation explanation is that people who have financial problems are more likely than average to turn to drugs (perhaps as a way to temporarily escape their financial worries). In this case, we'd consider the first explanation to be the more likely, but the reverse causation explanation is at least plausible. Indeed, both mechanisms could be operating simultaneously.

To test our hypothesis about tail streamer length, we are predicting that the length of a male's tail will affect whether he gets a lot of matings. But what if the number of matings a male gets affects his tail length; this would be a problem of reverse causation. Reverse causation would mean that although we think that the relationship between mating number and tail length shows that tail length affects mating success, we cannot discount the possibility that the reverse is true and mating success affects tail length (perhaps because mating induces release of hormones that affect feather growth).

Let's consider another example of reverse causation. There is a correlation between the number of storks nesting in the chimneys of a Dutch farmhouse and the number of children in the family living in the house. This sounds bizarre until you realize that larger families tend to live in larger houses with more chimneys available as nest sites for the storks. Large families lead to more opportunities for nesting storks; storks don't lead to large families!

Reverse causation is not a universal problem of all correlational studies. Sometimes reverse causation can be discounted as a plausible explanation. It would be extremely difficult to imagine how the number of matings a male gets later in a season could affect his tail length when we catch him at the beginning of the season. However, imagine that we had instead caught and measured the males at the end of the season. In that case it might be quite plausible that the number of matings a male gets affects his hormone levels. The change in hormone levels might then affect the way that the male's tail develops during the season in such a way that his tail length comes to depend on the number of matings that he gets. In that case we would observe the relationship between tail length and matings that we expect, but it would be nothing to do with the reason that we believe (females preferring long tails). Such a pathway might seem unlikely to you and you may know of no example of such a thing happening. However, you can be sure that the Devil's advocate will find such arguments highly compelling unless you can provide good evidence to the contrary. Experimental manipulation gets around the problems of reverse causation, because we manipulate the experimental variable.

These two problems of non-manipulative studies are captured in the warning, 'correlation does not imply causation' and are summarized in Figure 2.2. Many of the hypotheses that we will find ourselves testing as life scientists will be of the general form 'some factor A is affected by some factor B', and our task will be to find out if this is indeed true. If we do a correlational study and find a relationship between A and B, this provides some support for our hypothesis. However, the support is not in itself conclusive. Maybe the reality is that A affects B through reverse causation, or that some other factor (C) affects both A and B, even though A and B have no direct effect on each other. Such problems mean that, at best, our study can provide data that are *consistent* with our hypothesis. If we wish to provide unequivocal evidence that A affects B, then we must carry out a manipulative study. Thus, we would encourage you to think carefully before embarking on a solely correlational study.

Q 2.6 Would you opt for correlation or manipulation if you wanted to explore the following.
(a) Whether having a diet high in processed foods increases a person's propensity for depression.
(b) Whether a diet high in saturated fats increases the risk of a heart attack.

Third-variable (and to a lesser extent) reverse-causation effects are big drawbacks of correlational studies that are likely to make a manipulative study more attractive.

2.5.4 Situations where manipulation is impossible

In the last section we were pretty harsh about correlational studies. Yet many correlational studies are carried out every day, and the scientific journals are full of examples. There are a number of situations in which correlational studies can provide very valuable information.

Sometimes it is simply not possible to manipulate. This may be for practical or ethical reasons. You can imagine the (quite reasonable) outcry if, in an attempt to determine the risks of passive smoking to children, scientists placed a number of babies in incubators into which cigarette smoke was pumped for 8 hours a day.

It may also not be technically feasible to carry out the manipulation that we require in order to test our particular hypothesis. Modifying streamer

We see a relationship between A and B

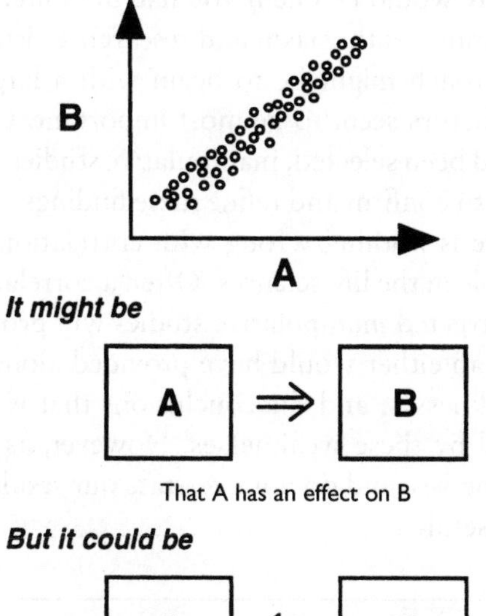

It might be

That A has an effect on B

But it could be

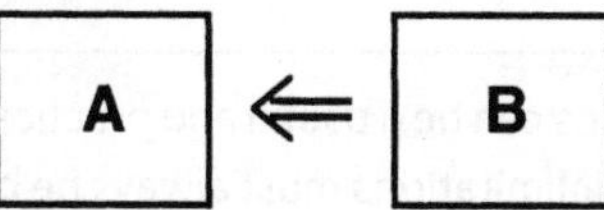

That cause and effect are the other way around and B has an effect on A. (This is reverse causation)

Or

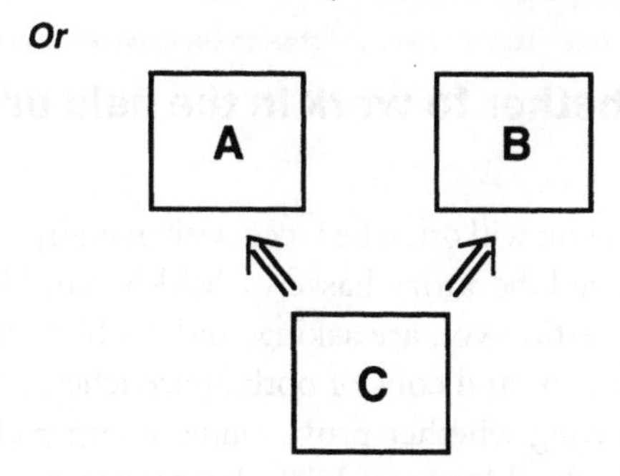

That A and B have absolutely no effect on each other, but both are affected by a third variable C

Figure 2.2 Schematic diagrams of the problems of reverse causation and third variables.

length was a difficult enough manipulation; experimentally modifying beak length or eye colour in a way that does not adversely affect the normal functioning of subject birds would be near-impossible.

A second valuable role for correlational studies is as a first step in a more detailed study. Imagine that you want to know what sorts of factors affect the diversity of plants in a given area. One possible approach would be to draw up a list of everything that we think might be important—all the soil chemicals and climactic factors—and then design a programme of experiments to systematically examine these in turn and then in combination. Such a study would be fine if you had unlimited research funds (not to mention unlimited enthusiasm and research assistants). However, a more efficient approach might be to begin with a large correlational study to see which factors seem to be most important. Once potentially influential factors had been selected, manipulative studies could be used in a more targeted way to confirm and refine these findings.

So ultimately there is nothing wrong with correlational studies; they fulfil an important role in the life sciences. Often a correlational study followed by carefully targeted manipulative studies will provide more compelling arguments than either would have provided alone. Correlational studies do have weaknesses, and the conclusions that we can draw will ultimately be limited by these weaknesses. However, as long as we are aware of these weaknesses, and do not overstate our results, correlational studies can be very useful.

Correlational studies can be a useful and practical way to address biological questions, but their limitations must always be borne in mind.

2.6 Deciding whether to work in the field or the laboratory

Another decision that you will often be faced with in many life science studies is whether it will be laboratory-based or field-based. Here the answer will depend on the question you are asking, and the biology of the system concerned. There are pros and cons of both approaches. Suppose that we are interested in knowing whether provisioning a larger clutch of young reduces a female zebra finch's survival. We decide that we will manipulate clutch size, by removing or adding eggs to zebra finch nests and then monitoring the survival of the mother after the chicks have fledged.

We could potentially carry out such an experiment in either the lab or the field. How should we choose? Let's begin with the lab. Maybe the most obvious question to ask is whether your study organism will be comfortable in the lab. In this case, we are unlikely to have a problem, as zebra finches will breed readily in captivity, but if we had been looking at the same question in a bird that is more challenging to keep in captivity (like an albatross), we might have had problems. The suitability of an organism for lab studies is hugely variable, and must be given careful thought. If an animal is not well disposed to captivity, then there are obvious welfare implications, as well as the risk that your results will have limited relevance to the situation in nature.

Now, assuming that your organism will be happy in the lab, there are several advantages of a lab study. In general it will be far easier to control conditions in a lab setting, allowing you to focus on the variable of interest without lots of variation due to other uncontrolled factors. Things like temperature, daylight, and the weather can all have profound effects on organisms, and variation due to these factors might mask the effects we are interested in. We can also ensure that all of our animals are well fed or our plants well watered, removing between-individual variation that could be caused by these factors in a natural setting. Observation will also generally be much easier in the lab. While the thought of watching zebra finches in the sunny Australian outback may sound idyllic, the reality of spending weeks trying to find your study birds is far less attractive. It will nearly always be easier to make detailed measurements on most organisms in the lab.

On the other hand, the controlled nature of the lab environment is also its major drawback. Laboratory individuals will not experience many of the stresses and strains of everyday life in the field, and that can make it difficult to extrapolate from lab results. Suppose that we find absolutely no effect of our laboratory manipulations on the lifespan of female zebra finches. What can we conclude? We can conclude that the manipulation had no measurable effect *under the particular laboratory conditions that we used*. This is very different from saying that it would have had no effect in the field, which is presumably what we are really interested in. It is quite conceivable that our well-fed laboratory birds in their temperature-controlled rooms and free from parasites can cope easily with extra chicks, but that in the harsher environment of the field they would have suffered a large cost. Of course, if we found no effect in the field, this result would only apply to the specific field conditions that occurred during our study, but those conditions are still likely to be more representative than the conditions in a laboratory. Thus, it is usually safer to generalize from a field study than a lab study. Generalization is a subject that is at the heart of statistical methodology, so we explore it a little further in Statistics Box 2.3.

Sometimes, our research question means that a laboratory study is impractical. If we are interested in differences between the sexes in parental care using zebra finches, then it would be very difficult to create conditions in an aviary that were close enough to those naturally experienced by wild birds. The danger is that any aviary experiments that you conducted on this would be discounted by Devil's advocates as being entirely artifactual, due to your asking birds to operate in a situation far removed from that which they have been evolutionarily adapted to cope with. However, if your research question is about how egg laying affects body reserves of calcium in females, then it would be easier to argue that this is less likely to be influenced by the husbandry conditions, and

STATISTICS BOX 2.3 **Generalizing from a study**

An important issue in the interpretation of a study is how widely applicable the results of the study are likely to be. That is, how safely can you **generalize** from the particular set of subjects used in your study to the wider world? Imagine that we measured representative samples of UK adult males and females and found that males in our sample were on average 5 cm taller. Neither we nor anyone else is particularly interested in our specific individuals; rather, we are interested in what we can conclude more widely from study of the sample. This becomes easy if we think about the population from which our sample was drawn. Our aim was to create a representative sample of UK adults. If we have done this well, then we should feel confident about generalizing our results to this wider population. Can we generalize even wider than this: say to European adults or human adults in general? We would urge a great deal of caution here. If your interest was in European adults, then you should have designed your study in the first place to obtain a representative sample of European rather than only UK adults. However, providing that you add an appropriate amount of caution to your interpretation, then it may be appropriate to use your knowledge of biology to explore how your results can be extrapolated more widely than your original target population. In this specific case, the specific difference of 5 cm would probably have little relevance beyond your specific target population of UK adults, since other countries will differ significantly in factors (e.g. racial mix, diet) that are highly likely to affect height.

hence a laboratory experiment might be reasonable. So, as ever, *the most appropriate approach depends on your research question.*

Our advice would be to aim to work in the field—generally giving results that are more easy to generalize from—unless it is impractical to do so.

2.7 *In vivo* versus *in vitro* studies

In many ways the question of lab or field has analogies in the biomedical sciences with whether to carry out studies *in vivo* or *in vitro*. If we want to know whether an anti-malarial drug alters the reproductive strategy of a malaria parasite, do we measure this using parasites in Petri dishes or within organisms? *In vitro* will generally be easier to do (although this is certainly not always the case), easier to control and easier to measure. But do the results have any bearing on what would happen in the natural system? The most effective approach will depend on the details of your study, and there are advantages and disadvantages of each. Indeed, the best study would probably contain evidence from both sources to give a more complete picture.

Issues of generalization of results will lead you towards *in vivo* experiments, whereas issues of practicality will lead you to *in vitro* studies.

2.8 There is no perfect study

Hopefully by now you will have got the impression that there is no perfect design that applies to all experiments. Instead, the best way to carry out one study will be very different from the best way to carry out another, and choosing the best experimental approach to test your chosen hypothesis will require an appreciation of both the biology of the system and the pros and cons of different types of studies. This is why *biological insight is a vital part of experimental design.* Let's illustrate these points by considering another example: the relationship between smoking and cancers.

In a sense, all of the data that we collect tell us something, but whether this something is useful and interesting can vary dramatically between data sets. Some data will be consistent with a given hypothesis, but will also be consistent with many other hypotheses too. This is typical of purely correlational studies, where it is difficult to discount the effects of reverse causation or third variables. A positive relationship between lung cancer and smoking definitely provides support for the idea that smoking increases the risk of lung cancer, but (as we discussed above) it is also consistent with many other hypotheses. Maybe it is due to stress levels that just happen to correlate with both smoking and cancer. Or maybe stress is an important factor, but there is *also* a direct link between smoking and cancer. In order to draw firm conclusions, we need something more. Our confidence might be improved by measuring some of the most likely third variables—like stress or social status—to allow us to discount or control for their effects. This would reduce the possibility that an observed relationship was due to some other factor, but would not remove it entirely (since there may still be some third factor that we've failed to account for).

If we were able to do an experiment where people were kept in a lab and were made to smoke or not smoke, any relationship between cancer and smoking would be far more convincing evidence that smoking causes cancer. If this could be repeated on people outside the lab, the evidence that smoking causes cancer in a real-world setting would become compelling. However, such experiments are ethically intolerable. Even if we can't do these experiments, we can do other things to increase our confidence. Maybe we can demonstrate a mechanism. If we could show in a Petri dish that the chemicals in cigarette smoke can cause cells to develop some pathology, this would increase our confidence that the correlational evidence supported the hypothesis that smoking increases the risk of cancer. Similarly, experiments on animals might increase our confidence. However, such experiments would entail a whole range of ethical questions, and the generality of the results to humans could be questioned.

Good experimental design is all about maximizing the amount of information that we can get, given the resources that we have available. Sometimes the best that we can do will be to produce data that provide weak evidence for our hypothesis. If that is the limit of our system, then we have to live with this limitation.

However, if what we can conclude has been limited not by how the natural world is but by our poor design, then we have wasted our time and probably someone's money too. More importantly, if our experiment has involved animals these will have suffered for nothing.

Hopefully, if you think carefully about the points in this chapter before carrying out your study, you should avoid many of the pitfalls.

There are always compromises involved in designing an experiment, but you must strive to get the best compromise you can.

Summary

- You cannot design a good experiment unless you have a clear scientific hypothesis that you can test.
- Pilot studies help you to form good hypotheses, and to test your techniques for data collection and analysis.
- Make sure that your experiment allows you to give the clearest and strongest evidence for or against the hypothesis.
- Make sure that you can interpret all possible outcomes of your experiment.
- Use indirect measures with care.
- Design experiments that give definitive answers that would convince even those not predisposed to believe that result.
- Correlational studies have the attraction of simplicity, but suffer from problems involving third-variable effects and reverse causation.

- Manipulative studies avoid the problems of correlational studies, but can be more complex, and sometimes impossible or unethical.
- Be careful in manipulative studies that your manipulation is biologically realistic and does not affect factors other than the ones that you intend.
- Often a combination of a correlational study followed by manipulation is very effective.
- The decision of whether to do animal experiments in the field or laboratory is determined by whether the test organism can reasonably be kept in the laboratory, whether the measurements that need to be taken can be recorded in the field, and how reasonably laboratory results can be extrapolated to the natural world. All of these considerations vary from experiment to experiment.
- There is no perfect experiment, but a little care can produce a good one instead of a bad one.

3 Between-individual variation, replication and sampling

- In any experiment, your experimental subjects will differ from one another.
- Experimental design is about removing or controlling for variation due to factors that we are not interested in (random variation or noise), so that we can see the effects of those factors that do interest us (section 3.1).
- One key aspect to coping with variation is to measure a number of different experimental subjects rather than just a single individual, in other words to replicate (section 3.2).
- Replication requires you to measure a sample of independent experimental subjects; section 3.3 is a warning about pseudoreplication, where your subjects are not independent of each other.
- Randomization is an effective means of controlling the effects of random variation (section 3.4).
- We finish in section 3.5 by advising you on how you might decide how many replicates you'll need in an experiment.

3.1 Between-individual variation

Wherever we look in the natural world, we see variation: salmon in a stream vary in their body size; bacteria in test tubes vary in their growth rates. In the life sciences, more than in physics and chemistry, variation is the rule, and the causes of variation are many and diverse. Some causes will be of interest to us: bacteria may vary in their growth rates because they are from different species or because a researcher has inserted a gene into some of them that allows them to utilize a novel type of sugar from their growth medium. Other causes of variation are not of interest; maybe the differences in growth rates are due to small unintentional differences in temperature between different growth chambers or to differences in the quality of the media that cannot be perfectly controlled by the experimenter. Some of the measured variation may not even be real, but due to the equipment that we use to measure growth rates not being 100% accurate (see Chapter 5). Hence we can divide variation into variation due to **factors** of interest, and **random variation** or **noise**. Of course, whether variation is regarded as due to a factor of interest or to random variation depends on the particular question being asked. If we are interested in whether different strains of a parasite cause different amounts of harm to hosts, the ages of individual hosts is something that we are not really interested in, but could nevertheless be responsible for some of the variation that we observe in the experiment. However, if we are interested in whether parasites cause more harm in older hosts than in younger hosts, the variation due to age becomes the focus of our investigation.

If we are interested in how one characteristic of our experimental subjects (variable *A*) is affected by two other characteristics (variables *B* and *C*), then *A* is often called the **response variable** or **dependent variable** and *B* and *C* are called **independent variables, independent factors** or sometimes just **factors**. It is likely that *A* will be affected by more than just the two factors *B* and *C* that we are interested in.

In very general terms, we can think of life scientists as trying to understand the variation that they see around them: how much of the variation between individuals is due to things that we can explain and are interested in, and how much is due to other sources. Whenever we carry out an experiment or observational study, we are either interested in measuring random variation, or (more often) trying to find ways to remove or reduce the effects of random variation, so that the effects that we care about can be seen more clearly. Life scientists have amassed a large array of tools to allow us to do this, and while the multitude of these and their specialist names can appear daunting, the ideas underlying them are very simple. In this chapter, we outline the general principles of experimental design, replication and randomization, and explain how these can be used to remove the masking effects of random variation. We also begin to think about the practical matter of how big a study needs to be. In Chapter 4 we deal with more specific techniques that can be used to increase the efficiency of our experiments. Together these two chapters cover the core of all good experiments.

Any variation in the response variable between individuals in our sample (between-individual variation) that cannot be attributed to the independent factors is called **random variation, inherent variation, background variation, extraneous variation, within-treatment variation** or **noise**.

Generally, we want to remove or control variation between experimental units due to factors that we are not interested in, to help us to see the effects of those factors that do interest us.

3.2 Replication

Random variation makes it difficult to draw general conclusions from single observations. Suppose that we are interested in whether male and female humans differ in height. That is, we want to test the hypothesis:

Gender has an effect on human height.

Replication involves making the same manipulations to and taking the same measurements on a number of different experimental subjects. Indeed, experimental subjects are often called **replicates** and their number the **number of replicates**. They may also be called **experimental units**.

To test our hypothesis we might find the heights of Pierre and Marie Curie from historical records, and find that Marie was shorter than Pierre. Does this allow us to conclude that human females are shorter than males? Of course not! It is true that some of the difference in height between Pierre and Marie will be due to any general effects of gender on height. However, there will be a number of other factors that affect their height that are nothing to do with gender. Maybe Pierre had a better diet as a child, or came from a family of very tall people. There are numerous things that could have affected the heights of these two people that are nothing to do with the fact that Pierre was male and Marie female. The difference in height that we observe between Pierre and Marie might be due to the factor that we are interested in (gender), but could equally just be due to other factors.

The solution to this problem is to sample more individuals so that we have replicate males and females. Suppose we now go and measure ten married couples and find that in each case the man is taller than the woman; our confidence that men really are taller than women on average would increase. The reason for our increased confidence is not very profound. Common sense tells us that, while it is easily possible that differences in height between a single male and a single female could be due to chance, it is very unlikely that chance will lead us to select ten out of ten couples where the man was taller, when actually couples where the woman is taller are just as common in the population. This is as likely to occur by chance as tossing a coin ten times and getting ten heads. It is unlikely that in all the chosen couples the males had better diets as children than the females, or all the females came from characteristically short families (unlikely but not impossible; see section 1.4.2 and Chapter 4 for more consideration of such confounding factors). If we found the same pattern with 100 couples we would be even more confident. What we have done is **replicate** our observation. If differences were due only to chance, we would not expect the same trend to occur over a large sample, but if they are due to real differences between males and females we would expect to see consistent effects between replicate measurements. The more times we make a consistent observation, the more likely it is that we are observing a real pattern. All statistics are based on replication, and are really just a way of formalizing the idea that the more times we observe a phenomenon the less likely it is to be occurring simply by chance. See Box 3.1 for a much more detailed treatment of the relationship between random variation and replication.

Replication is a way of dealing with the between-individual variation due to the random variation that will be present in any life science experiment. The more replicates we have, the greater the confidence we have that any difference we see between our experimental groups is due to the factors that we are interested in and not due to chance.

Q 3.1 Are we safe in restricting our sample to married couples?

Replication lets us look for consistent patterns that are unlikely to be explained by chance.

BOX 3.1 A detailed example of variation and replication: do seed beetle larvae experience competition?

Given the importance of variation and replication to good experimental design, it is worth going into these ideas in a little more detail using an example: *Callosobruchus* beetle larvae that develop inside legume seeds.

First imagine an idealized situation where competition was the only thing that affects beetle size

Imagine instead of the real world, we lived in a world without random variation between individuals, where the only cause of variation in adult beetle size was whether or not an individual suffered from competition as a larva. In this imaginary world, all beetles that share beans as larvae emerge with a weight of exactly 5 mg, while all those that develop alone develop at exactly 6 mg. A scientist that was lucky enough to live in this world would have no need for either replication or statistics. In order to quantify the effect of competition, they could measure a single beetle raised in a bean with another larva and a single beetle raised alone, and compare the two measurements. They are seeking to test the hypothesis:

> A beetle's weight is influenced by whether or not it competed with another individual during the larval stage within a bean.

What would be the point of replication when all the replicates that grew up alone would weigh exactly the same (6 mg), as would all the beetles that experienced competition (all 5 mg this time)?

Getting back to the real world where other factors affect size

In the real world, the size of a beetle may be affected by whether it is raised in competition or not, but also by a number of other factors that vary between beetles. These might include differences in the quality of the bean that they develop in, variation in temperature and humidity, and many other factors. These other factors will lead to variation between individual beetles in the two groups ('raised alone' and 'raised with another'). If we are only interested in studying the effect of competition on adult size, then this added variation (due to bean quality, for example) is just a nuisance that makes it harder for us to study the effect that we are interested in.

We can think of the weight of a beetle as being made up of two parts. The first part is a baseline that is determined by whether the beetle has experienced competition. Beetles raised together have a baseline of 5 mg, and beetles raised alone one of 6 mg. However, the effects of all the other factors that influence an individual's weight mean that few beetles will actually weigh exactly 5 or 6 mg. Some will have been lucky; the bean they develop in might be of slightly better quality, or the humidity might be slightly more suitable for larval development. These beetles will emerge slightly heavier than their baseline. Other beetles will be less fortunate, and the particular values of the random factors that they experience will mean that they end up smaller than the baseline. Thus the actual size of a beetle can be thought of as being the baseline for the particular group that it is in (i.e. 'raised alone' or 'raised with another') plus a deviation due to the

BOX 3.1 ***(Continued)***

particular set of all the other factors that it experiences. In the real world we get a distribution of masses from beetles that experienced competition and another distribution of masses from beetles that did not, as illustrated in Figure 3.1(a).

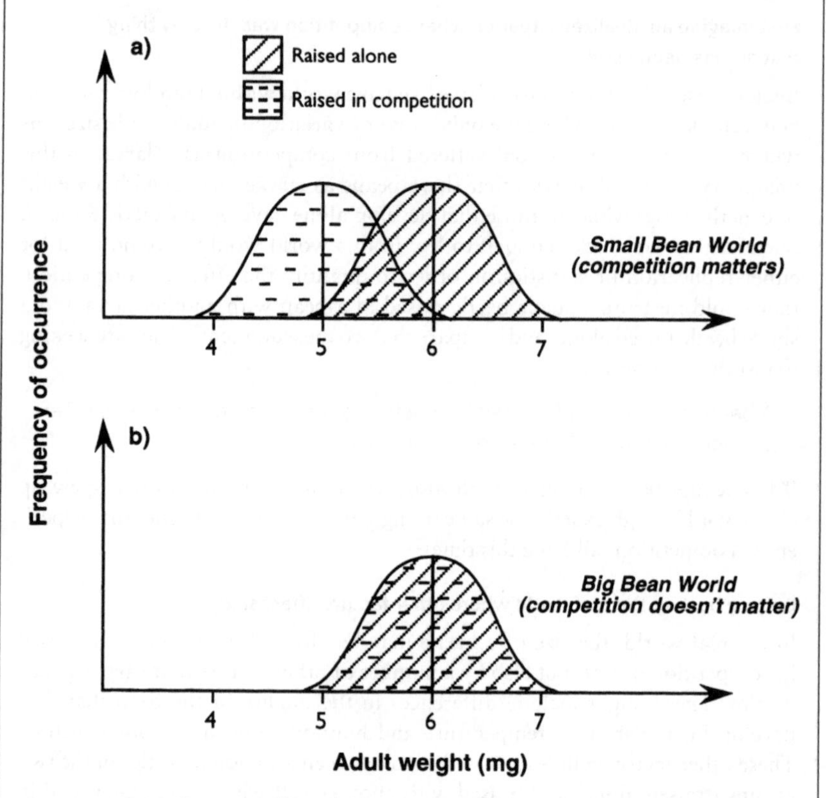

Figure 3.1 Distributions of beetle weights.

The **standard deviation** is a measure of the spread of values around the mean (or average) value. If the distribution of values is a Gaussian shape, then 95% of the values are within two standard deviations of the mean. So if the mean weight of beetles raised alone is 6 mg and the standard deviation is 0.5 mg, then 95% of the beetles raised alone weigh between 5 and 7 mg.

In the Figure we can see that the beetles raised without competition cluster around 6 mg, but many are slightly bigger or smaller than this, and a few are much bigger or smaller than this. This situation leads to a bell-shaped distribution of beetle sizes. Technically, this distribution is called a normal or Gaussian distribution, and this type of distribution crops up throughout biology. How spread out a normal distribution is depends on how much variation due to random factors there is, and this is measured by the **standard deviation** of the distribution. If there is a lot of random variation, the distribution will have a high standard deviation, and be very widely spread. In our case that would mean that the random factors would often cause beetles to be much bigger or smaller than their baseline. Alternatively, if there is not much random variation, the distributions will be narrow and have a small standard deviation, and so

BOX 3.1 (***Continued***)

most beetles will be very close to their baseline. The standard deviation of the distributions in Figure 3(a) is 0.5 mg. The beetles raised with competition also form a normal distribution with a standard deviation of 0.5 mg, but because competition reduces their baseline, their distribution has a mean of 5 mg.

Now let's describe a different world, an alternative world where competition has no effect on beetle size. Such a world is summarized in Figure 3.1(b). Again, we can think of there being two distributions of beetle sizes, for beetles raised with and without competition. However, in this world competition has no effect, so the baselines of the two groups are the same, and the two normal distributions have the same mean, i.e. the average beetle size is the same in both populations. To simplify things, we have also assumed that their standard deviations are the same.

How can we find out which model of reality is correct?

Let's refer to these two different worlds as Small Bean World (where sharing a bean adversely affects a beetle's weight) and Big Bean World (where sharing a bean with another beetle has no effect on a beetle's weight). Now imagine a researcher who wants to investigate whether competition affects size in beetles. The task facing the researcher is to decide whether they live in Small Bean World or Big Bean World. Suppose that the researcher measures two beetles, one raised in competition and one raised without competition, and finds that the beetle raised in competition weighs 5.1 mg, while the beetle raised without competition weighs 6.1 mg. Does this show that competition reduces the size of a beetle? Well, it is certainly consistent with the hypothesis, but it is not really compelling evidence. To see why, let's go back to the pictures of the two alternative worlds. Doing the experiment is like picking one beetle at random from each distribution. If we do this in Small Bean World, we are quite likely to pick a beetle from a shared bean that is 1 mg smaller than a randomly selected beetle that was alone in a bean, because competition really reduces the size of beetles. However, even in Big Bean World we are quite likely to pick beetles that differ in weight by at least this much, even though competition has no effect and beetles differ in weight only through random variation. Thus, with a single observation like this, the investigator has no way of telling these two worlds apart.

The key is to replicate

The researcher decides to repeat the experiment, but this time to measure two beetles of each type. The means of the two sets of beetles are (coincidentally) 5.1 and 6.1 mg for beetles raised with another and raised alone respectively. The researcher can now begin to be a bit more confident that competition is having an effect. While in Big Bean World the chance of a beetle raised with another being 1 mg smaller than a beetle raised alone is quite high, the chance of the mean of two beetles (each raised with another) being that much smaller than the average of two beetles raised alone is lower. This is because whether a beetle is big or small in Big Bean World is only a matter of chance. For the mean of two beetles to be big, generally this requires that both the beetles are big, which is

BOX 3.1 (***Continued***)

less likely to happen by chance than if only single beetles were selected. However, even with two replicates it is still quite possible to get this result in Big Bean World, just due to chance. The only thing to do to improve our ability to tell the two worlds apart is to take more replicates.

With five beetles in each group, the chance of getting means for the two types of beetles that differ by at least 1 mg in Big Bean world is becoming quite small. It would require that by chance we had picked several beetles raised with another that were atypically small, and several raised alone that were atypically big. So with five beetles per group we can be more confident that such a big difference in the mean weights between groups is very unlikely to be due to chance. With ten beetles per group the probability becomes smaller still.

Although in Small Bean World we know that competition reduces a beetle's size by 1 mg, with only one measurement per group, chance will often lead to the two sampled beetles differing by less than this, and sometimes even to a beetle raised with another being bigger than a beetle raised alone. As the sample size increases, the chance that we will see a difference of 1 mg between the average weights of the two groups increases.

In a sense we can think of replication as causing the effects of random variation to cancel out. As the sample sizes increase, the effect of random variation is reduced, and so any differences we observe are more likely to be due to the factor that we are interested in, rather than the many other factors that cause variation between individuals.

Replication and statistics

Sometimes, even if we are in Small Bean World, we will find that our group of beetles from the 'raised with another' treatment is not much smaller than that from the 'raised alone' treatment. Sometimes they might even be bigger, just through chance. And sometimes a researcher in Big Bean World would find large differences between the means of large groups of beetles. However, if we have large enough groups, and we find big enough differences between their means, we can say that the populations from which the samples are drawn differ fundamentally and be confident that we will be right most of the time. This is the basis of all statistical tests. Statisticians have devoted many years to developing tests that allow us to decide whether the patterns that we observe in our experiments (differences between means, relationships between variables) are probably because of real biological effects, or could be just due to chance. These tests rely on the amount of variation that is measured, and the differences that are seen, and they have many different names and specific assumptions. However, underlying all of these tests is the same logic as was discussed with the beetles above; they allow us to look at the results of our experiment and ask: what is the likelihood that these results represent a pattern that was not generated simply by chance? The bigger the difference between the groups that is observed, and the bigger the sample sizes, the more likely it is that the effect is real and not just due to chance. However, it is always worth remembering that statistics deals in probabilities, it never deals in certainties.

3.3 Pseudoreplication

3.3.1 Explaining what pseudoreplication is

As we have just seen in the last section, replication allows us to deal with the between-individual variation that is due to chance effects, and so see any biologically interesting patterns more clearly. However, this effect of replication relies on one critical rule: the replicate measures must be independent of each other. What this means is that any measurement is just as likely to have a positive deviation from the norm due to random variation as it is to have a negative one. If this is the case, this will have the important effect of meaning that if we examine a group of independent individuals their deviations will tend to cancel out, and the mean of the **sample** will be close to the mean of the **population**. Technically, statisticians talk about there being no correlation between the deviations of individuals within a group. However, if we don't design our experiments carefully, this will not always be the case.

We are interested in asking questions about a certain group of individuals (the **population**). Generally this population is too large for us to study every individual, so we study a representative subset (the **sample**).

In the life sciences, we are almost always working with a sample of experimental subjects, drawn from larger population. To be independent, a measurement made on one individual should not provide any useful information about that factor in another individual. For example, imagine that we wished to test the hypothesis:

> Blue tit nestlings raised in nest boxes suffer more from external parasites than those raised in natural cavities.

If we investigate the four nestlings in a particular nest box and count the number of parasites on each, then this does not give us four independent measures of the parasite load of nest box chicks. The problem here is that there is highly likely to be a strong correlation between the parasite loads between nest-mates, so if we find that one has a high parasite burden then it is likely that all the others do too. In this experiment broods rather than nestlings should be considered as the experimental unit, and for each brood we should perhaps record the mean number of parasites per nestling, calculated over all nestlings of that brood.

Let us go back to the heights of male and female humans. Suppose that, rather than measuring a man and his wife once each, we measure each of them ten times. We now have replicate measurements; can we now draw reliable conclusions about the heights of males and females? Again, the obvious answer is no. The ten male measurements are not independent of each other: they were made on the same man. All the measurements on the man will have the same baseline height due to 'maleness', but they will also have the same values for all the other factors that affect the height of this one man. Every measurement of the man's height will be affected in the same way by his diet as a child, the size of his parents, and every other factor that influenced his height. Thus, if our first measurement of the man's height had a positive deviation from the norm (i.e. he was taller than the average height for men), all of the other measurements we make on him will also have positive deviations, and so the deviations will not tend to cancel out. Statisticians would say that the deviation of one measurement of the man's height was correlated with the deviation of another measurement of his height. If the measurements were independent measures of males and females, then the only consistent difference would be due to gender. The fact that there are a whole number of other factors that differ consistently between the replicate measurements of the one man and those of his wife mean that they cannot be independent measures of the effect of being male or female on a person's height. Statisticians would say that the differences due to gender are confounded with all the other things that differ between the two specific individuals that we measured. Hence we do not have ten replicates of male height, but ten pseudoreplicates. If we treated these measurements as independent replicates, we would be guilty of pseudoreplication.

More generally, if we are interested in the possible effects of factors B and C on A, but factors B and C are closely related such that B strongly influences C, then we cannot tell if A is affected by B alone, by C alone or by both factors, since B and C are confounded. For example, it can be difficult to evaluate whether egg production increases with age or experience in a population where all individuals start breeding at the same time and breed continuously thereafter, since age and experience are correlated and therefore confounded: all the older individuals are experienced and all the young individuals are inexperienced.

Failure to have independent replicates is a very serious problem for an experimental design, as almost all statistical tests demand independence. But surely no one would ever make this sort of mistake? In fact, this sort of error is made all the time, and one of the most common places to see it is in the media.

Consider the following 'experiment' that was carried out by a British national newspaper. The experiment was to test whether different groups in society were treated differently when they complained to organizations about poor service that they had received. The newspaper selected eight people, including a student, a vicar, a disabled person, and an unemployed person, and asked them to write letters of complaint to six different companies. They then monitored the replies that each person got and concluded that students, handicapped and unemployed people were treated far worse than the others. Was this conclusion justified? At first sight it seems that they have carried out a replicated study; they have six replies for each type of person. However, rather than being true replicates, these are pseudoreplicates, because all the student measurements came from a single student, all the vicar ones came from a single vicar, and so on. The responses that the vicar received could have been partly due to any effect of being a vicar (the effect that the study was interested in), but could also be due to any number of other factors that were nothing to do with being a vicar. Maybe this particular vicar had neater handwriting than the particular student used. The differences between organizations' reactions to complaints by vicars and students are confounded with all of the other differences between the two individuals concerned, and so tell us nothing about the general differences between vicars and students.

A better way to conduct this survey would have been to compose a standard letter and send multiple copies of it, differing only in the stated occupation of the putative author. This would remove individual variation in phrasing as a confounder. If these letters were produced on a word processor, then this would remove handwriting style as a confounder. We might also want to control for postmark by posting them all from the same place. We would have to send them at different times to the same company, or the person receiving eight identical letters from allegedly different people would find it rather strange. We would want to make sure that the order in which the letters were sent varied between companies. If the student letter was always sent before the vicar letter, then this would introduce time as a confounder. From this example, you can see that pseudoreplication is a trap that it can be easy to fall into.

Replicates must be independent: if they are not, then they are pseudoreplicates.

3.3.2 Common sources of pseudoreplication

The use of multiple measurements from the same individual as if they were independent measurements is an extreme and obvious form of pseudoreplication, and it would be unusual, but sadly not unheard of, to see such mistakes being made in life science research. However, biology is full of many other sources of non-independence that can catch out the unwitting researcher. Here are a few to watch out for.

A shared enclosure

Imagine that we want to know whether adding enrichment to primate enclosures decreases the level of aggression by the inhabitants. We go to our local zoo, where they have two enclosures, each containing ten chimps. We add a selection of toys and playthings to one enclosure in an attempt to enrich the enclosure. We leave the other enclosure unmodified.

If we then observed the two enclosures and found that the level of aggression of the ten chimps in the enriched enclosure was significantly lower than that of the ten in the other enclosure, is this good evidence that enrichment reduces aggression? Unfortunately it is not, because the behaviours of the ten chimps are not ten independent measures of the effect of enrichment. The problem is that there will be many other differences between the two enclosures; maybe one is warmer or sunnier than the other, or slightly bigger than the other. Thus chimps in an enclosure share many other factors that are nothing to do with the enrichment, and any of these factors could explain the difference in behaviour between the groups.

This type of problem can crop up in any number of situations. If your experiment looks at the effect of different temperatures on the courtship behaviour of crickets, but all the crickets for one temperature treatment are in one incubator and all the crickets for the other are in another, then the crickets are not independent measures of the effect of temperature, because temperature is confounded by all the other differences between the two incubators.

The common environment

A similar source of pseudoreplication is a common environment. Suppose that you are interested in whether red deer that feed in woodland have higher parasitic worm burdens than those that feed on moorland. You take a train to the highlands, find a wood and a moor where red deer are feeding, and collect faecal samples from twenty deer at both sites and determine the number of nematode eggs in the samples. Does this allow you to see whether woodland deer have higher worm burdens than moorland deer? While the most obvious difference between the two sampling sites will be the vegetation, there will be many other differences too (perhaps the wooded site is at higher altitude, for example), and any of these differences might affect worm burdens. Thus the deer from each site share many other factors that are nothing to do with the presence or absence of woodland at the site, and the deer are pseudoreplicates of the effect of woodland.

Relatedness

We all know that genetics mean that relatives tend to be more similar to each other than do unrelated members of a population. This similarity due to genetics means that relatives are not independent data points when looking at the effects of other treatments. So if we were looking at the effects of fertilizers on seed development, but all of the seeds in one treatment group came from one adult plant, while all the seeds in the other treatment came from another adult plant, these seeds would not be independent measures of the effect of the fertilizer treatment. Differences between the two groups of seeds might be due to the fertilizer treatment, but could equally be due to any genetic differences between the two groups. In the studies of deer and chimps discussed above, as well as the problems highlighted, there are also possible problems of relatedness between the individuals in each chimp enclosure or at each deer site.

A pseudoreplicated stimulus

Imagine that we were interested in whether female zebra finches prefer males with bright red beaks. We might carry out the following type of experiment. We take two stuffed male zebra finches and place them on opposite sides of a cage. One of the males has a bright beak, the other a dull beak. We then place twenty females in the cage one at a time, and see which model they spend most of the time next to. We find that, on average, females spend more time with the bright male than the duller male. Can we now conclude that females prefer males with bright beaks? Again, the answer is 'no', and the reason is that our stimuli are pseudoreplicated. It is true that the stuffed finches differ in their beak colour, but they probably differ in any number of other ways as well; for example, maybe one is slightly bigger than the other. The differences that we observe in response to the two models might be due to their beak colours, but might be due to any of these other differences. Multiple measurements using the same two models are no more independent measurements of the effect of beak colour than multiple measurements of Pierre and Marie Curie are independent measurements of the height of male and female humans.

Q 3.2 Why in the above mate-choice experiment do we test the females individually, rather than saving time by introducing them all to the choice chamber simultaneously?

Individuals are part of the environment too

A particular problem of behavioural studies is that individuals will affect each other. The chimps that we discussed above are social animals, and the way in which one chimp behaves will have effects on the other chimps in the enclosure. So a single aggressive individual in a group might make all the others aggressive. This can be thought of as an extension of the shared environment problem, but with the animals being the part of the environment that is shared. The problem is not limited to behaviour either, but will arise whenever the state of one individual in a group can affect the state of others. With the deer and worms discussed above, it is quite possible that if one individual in a group has lots of worms, it will infect all the other members of its group, and so the worm burdens of the members of a group are not independent.

Pseudoreplication of measurements through time

Imagine you are interested in the ranging behaviour of elk. Specifically, you want to know whether a particular elk individual prefers different types of habitat. To carry out your study you manage to get hold of some satellite tracking equipment that can tell you the position of the elk at specified time intervals. For each measurement that you make, you note down the type of vegetation that the elk is in. You find that of the twenty measurements that you made, nineteen occurred in habitat that you classified as dense woodland, and only one occurred in open woodland. In your survey area as a whole there are equal amounts of these two habitat types, so does this evidence suggest that this elk prefers to spend time in dense woodland? The validity of this conclusion will depend on whether each of your measurements on an individual elk can be regarded as independent measures of the position of that elk. Whether or not this is true will depend (in part) on how much time has passed between measurements. Imagine first that we have taken the position of the elk once a day (at a random time of day) for 20 days, and that our study area is such that an elk could easily walk from any point in the site to any other point in the time between measurements. If that is the case, then we have no reason to expect that the position of an elk at one measurement point will affect its position at any other measurement point. Contrast this with the same study, but with measurements taken every second for twenty consecutive seconds. In this case the position of an elk at one measurement point will obviously have an effect on where it is found at the next measurement (1 second later). An elk can obviously not move far in a second, and so an elk that is found in dense woodland at one point is far more likely to also be in dense woodland a second later. If we had been silly enough to carry out the study in this way, we would not have independent data points, and our inference about habitat use would be flawed. Because of the non-independence of data points, even an elk with no habitat preference would still be highly likely to be found in the same habitat type every second for 20 seconds. Remember also that even if we collect good data of the ranging behaviour of this elk, it is only one individual, and it would be risky to assume that it is a representative individual of the wider population; we would want to replicate measurements with different elk in order to draw conclusions about more than one individual.

Similar problems can occur in any experiment where we are taking multiple measurements through time, and whether or not measurements are independent will depend critically on the biology of the system. Even daily measurements might not be independent if we were studying a snail rather than an elk in the example above. Once again, you can see that deciding whether you have a pseudoreplication problem is a decision that involves understanding of the biology of the study system.

Species comparisons and pseudoreplication

It is often not possible to directly manipulate traits experimentally. However, as life scientists, we are fortunate to have a wide variety of species with different trait values that allow us to make comparisons. Imagine we hypothesize that, in species where females mate with several males, sperm competition will lead to an increase in the relative testes size of the males. To test this, we take two species of damselfly, one where females mate once, and another where females mate multiply. We then compare the relative size of the males' testes, and find that those of the males in the promiscuous species are bigger for their body size than those of the monogamous species. Can we conclude that our hypothesis is correct? Well not really, because the measurements on the individuals of the two species are not really independent measurements of the effect of mating system. It is true that all the males of a species do experience the same mating system and this differs between the species, but they all experience a whole range of other factors that differ systematically between the species as well. The differences observed might be due to the factor that we are interested in, but might be due to any of the other factors that differ consistently between species. Thus multiple measurements of a species are not independent measures of any single difference between species.

Q 3.3 You are engaged in testing whether two factors vary between men and women: propensity to have red hair and willingness to engage in conversation with strangers. While waiting in the dentist's waiting room to be called through for a check-up, you realize that the other occupants of the waiting room might provide an opportunity to add to your data-set. How many independent data points can you collect?

So what about if we look at several species of damselfly and find that the pattern is consistent? Can we now have confidence that the effect is due to mating system? Unfortunately the situation is not that simple. The dynamics of the evolutionary process means that closely related species will tend to be similar to each other simply because of their shared evolutionary history. Thus suppose we have ten species of promiscuous damselfly, all with relatively large testes, but that these have all evolved from a single ancestral species that was promiscuous and had large testes. We would have to be very cautious about drawing strong conclusions about our hypothesis based on our ten species. Furthermore, closely related species are likely to share many other aspects of their ecology too. There are statistical methods developed to deal with these problems, and Harvey and Purvis (1991) in the bibliography is a good place to start if this interests you.

It's easy to pseudoreplicate if you are unwary.

3.3.3 Dealing with pseudoreplication

After the last section, you must be thinking that pseudoreplication is a minefield that will cause you no end of trouble. Not so; forewarned is forearmed, and you will avoid lots of pitfalls just by bearing in mind the cautions of the last section. However, this section will offer more practical advice, armed with which pseudoreplication should hold no fears.

Pseudoreplication is a biological issue

The most difficult thing about pseudoreplication is that there are no hard and fast rules. Instead, whether things are pseudoreplicates will depend on the biology of the species that you are studying and the questions that you are asking. *Thus pseudoreplication is a problem that has to be addressed primarily by biologists and not by statisticians.* If you give a data set to a statistician, there is no way that they will be able to tell by looking at the data if your replicates are independent or not. A statistician might be able to guess that the behaviour of one chimp will affect the behaviour of others in the same social group, but can't be expected to know this, let alone whether the behaviour of beetles that are kept in separate pots inside an incubator are likely to affect each other. This is why biologists need to be involved in the design of their experiments. So use your biological knowledge to ask whether two measurements are independent or not.

There are many solutions to pseudoreplication, and the course of action that you take to avoid it will depend on the particular situation that you are in. Sometimes the solution is straightforward: in the female choice experiment with the pseudoreplicated stimulus, one solution might be to sample from a selection of stuffed males so that every female is tested with a different pair of stuffed males.

A more general solution involves condensing and replicating. In the example of the deer, we decided that individual deer were not independent measures of woodland or moorland habitats. The first step is to condense all the measures of the woodland deer to a single data point. This might be achieved by picking one at random, or by taking the mean for all the deer at the woodland site. We then do the same for the moorland site. However, this still leaves us with a problem, as we now have only one measurement for woodlands and one for moorland. All the replication that we thought we had has vanished. The solution is then to go and find more woodlands and moorlands and do the same at each. This would give us several independent data points of deer from woodland sites and several of deer from moorland sites, and any differences we found could then be put down to more general effects of woodland or moorland. The same procedure could be used for the insects in the incubators (take a mean for the incubator, and replicate incubators), and the chimps in the enclosures (take a mean for the enclosure, and find more enclosures). Throwing away all that data by lumping the measurements all into one mean value might seem like a terrible waste of effort. However, it would be far more of a waste of effort to pretend that the measurements really were independent and draw unreliable conclusions. It is worth asking at this point whether this implies that measuring several deer at each site was a complete waste of time. Would we have been just as well measuring only a single deer at each site, given that we ended up with a single measurement at each site? This is a topic that will be discussed in more detail in section 6.2. Here we will simply say that because the mean of several deer at a site will be less affected by random variation than a single measurement, it will sometimes pay to measure several animals at a site, even if we ultimately condense these into a single mean. The extra work of collecting measurements from more deer at each site will give a more accurate measure of that site, and mean that fewer different sites will be required in your final analysis. As a life scientist, you'll probably enjoy looking for deer dropping more than driving between sites, so this is an attractive trade-off.

Sometimes it won't be possible to replicate in this way. Maybe the lab you are working in only has two incubators. Does this mean that your study of the effects of temperature on courtship song in crickets is doomed? There is another type of solution that is open to you in this situation. Imagine that we run the experiment with incubator A at 25°C and incubator B at 30°C. We can then repeat the experiment with incubator B at 25°C and incubator A at 30°C. We could do the same with the enclosures in the enrichment example, first enriching one enclosure, then the other. This means that we now have replicate measurements of crickets raised in incubator A at both temperatures, and also in incubator B at both temperatures. By comparing the beetles at different temperatures in the same incubator, we have independent measures of temperature effects (although even this assumes that the incubators do not change between our runs of the experiment). This is an example of a statistical technique called blocking, which will be discussed in more detail in the next chapter (section 4.3).

Q 3.4 You are carrying out a clinical trial of a drug that is reputed to affect red blood cell count. You have 60 volunteers that you have allocated to two groups of 30. One group is given the drug for 1 week, the other is not. You then take two blood samples from each person, and for each blood sample you examine three blood smears under a microscope and count the red blood cells. This gives you a total of 360 red blood cell counts with which to test your hypothesis. Would you be concerned about pseudoreplication in this study? If so, how would you deal with the problems?

Q 3.5 Was measuring six smears from each individual in Q 3.4 a waste of time?

3.3.4 Accepting that sometimes pseudoreplication is unavoidable

But what if we cannot condense data effectively and we cannot block: is all lost? Suppose that we really can only survey the deer on one woodland and one moorland—does this mean that our study of the effect of woodland on parasitism is a waste of time? Not necessarily. The important thing is to ask yourself what the data tell you. Suppose we did find more worms in the woodland deer than the moorland deer; what can we infer? Well, we can say with some confidence that groups of deer differ in their worm burdens, but this is unlikely to be a very exciting result. If the deer at the two sites are similar (in species, density, age structure, etc.), then we can have some confidence that this is due to differences between the two sites rather than other differences between the two groups of deer. However, we can't say that this is due to the fact that one site is woodland and one site grassland, because of all the other differences between the two sites. So what began as a study of an interesting question 'do grassland deer have fewer worms than woodland deer?' has been limited to answering the question, 'do all groups of deer have the same worm burdens?'. This is typical of what happens if we don't think carefully about our replicates: we think we are answering an interesting question, but in fact we can only answer a much more mundane question, to which we probably already know the answer. Replicates allow us to answer general questions, and the generality of the answer that we get will depend critically on how we sample and replicate. If we are interested in the effect of gender on the heights of humans, multiple measurements of Pierre and Marie Curie are no use. They tell us about differences between Pierre and Marie, and if we can only measure their heights with a great deal of error we might need multiple measurements even to answer such a simple question. However, these multiple measurements will tell us very little about gender differences. If we measure the heights of several males and females, then, as long as we have avoided traps of relatedness, shared environments (did all the males come from one country and all the females from another?) and so on, we can begin to say something more general about the effect of gender on height. However, even if we have sampled carefully, if all the individuals come from Britain, they are independent measures of the heights of British males and females, but not of males and females at the global level. So if we were interested in knowing the answer to this question at a global level, we would need a sample of independent replicates drawn from the global population.

So the key if you can't replicate fully is to be aware of the limitations of what you can conclude from your data. Pseudoreplicated enclosures mean that we can talk about whether there are differences between the enclosures, but we can't be sure what causes any differences, and pseudoreplicated stimuli mean that we can talk about the effects of the models on female choice, but can't be sure that these effects are due to the beak colour that we were originally interested in.

3.3.5 Pseudoreplication, third variables and confounding variables

In the previous chapter we introduced the problems of third variables in relation to correlational studies, and in this chapter we have talked about replication, pseudoreplication and confounding variables. Here we will discuss how these ideas are linked.

As we have already said, if we want to know the effect of some experimental treatment, say the effect of an antibiotic on bacterial growth rates, we need independent replicate measures of the effect of that treatment. That is, we need several measures of the growth rate of bacteria with and without the antibiotic, and the only systematic difference between the bacteria in the treatment and control groups must be the presence of the antibiotic. If the bacteria in the two groups differ systematically in other ways as well, then our replicates are actually pseudoreplicates. For example, imagine we took two agar plates containing bacterial growth medium, and added antibiotic to one of the plates. We then spread a dilute solution of bacteria on each plate, and measured the diameter of

100 bacterial colonies on each plate after 24 hours. We appear to have 100 measures of growth rate with and without antibiotic, but of course these are not independent measures, because all the bacteria in a group are grown on the same plate. The plates differ with respect to antibiotic application, but they will also differ in other ways as well; the differences due to antibiotic will be confounded with these other differences. Or, to put this in the language of Chapter 2, we cannot be sure that the difference we see is due to the factor that we are interested in or to some unmeasured third variables. So confounding variables and third variables are the same thing, and if they are not dealt with appropriately they will turn our replicates into pseudoreplicates. One solution to the problem in the bacterial study might be to have several plates with and without antibiotic, and then measure the growth rate of a single colony on each plate (or of several colonies and then take a mean for each plate).

To think about it in another way, the problems of third variables in correlation studies can equally be thought of as a problem of pseudoreplication. Because we have not manipulated experimentally, we cannot be sure that the individuals that we are comparing really are replicate measures of the effects of the factor that we care about, rather than being pseudoreplicates, confounded by other factors.

3.3.6 Cohort effects, confounding variables and cross-sectional studies

Another area where non independence and confounding effects can trip up the unwary is when we are interested in changes that occur through time. So, imagine we find that a sample of 40-year-old men had better hand–eye co-ordination in a test situation than a sample of 20-year-old men? Can we safely assume that men's co-ordination improves as they get older?

Our study is cross-sectional, where we take a snapshot in time and compare subjects of different ages. If we find a difference between different age classes, then this could indicate that individuals do change over time, but there is an alternative explanation. It could be that individuals do not change over time, but that there is a difference between the individuals in the different age classes due to some other factor. For example, the types of jobs that men typically do has changed dramatically over the last 20 years. Perhaps a greater fraction of the 40-year-olds have jobs that involve manual labour or craftsmanship, whereas more of the 20-year-olds have office jobs. If the type of job that a person does affects their co-ordination, then this difference in employment may explain the difference between the groups. You will recognize this immediately as another example of a confounding factor, although in this type of study it generally goes by the name of a cohort effect. An alternative way to probe the question of whether an individual's co-ordination changes with age is to follow individuals over time and compare performance of the same individuals at different ages: this would be called a longitudinal study. But, hang on, given what we have said in section 3.3.1 about multiple measurements on the same individual not being independent, alarm bells should now be ringing. Surely, by measuring the same individuals at different ages, we are doing just that? For the time being you will just have to take our word for it that in situations like this, where we are specifically interested in how individuals change with time or experience, then (as long as the data are handled with care) this is not a problem. The details of why it is not a problem can be found when we discuss within-subject designs in section 4.4. Thus a longitudinal study avoids the potential for being misled by cohort effects, but obviously takes longer to provide us with an answer than a cross-sectional study. Hence a reasonable compromise is often to take a combination of the two.

As another example, imagine that we wanted to explore fecundity in female lions to see if it declines after a certain age. If a cross-sectional study showed that 3-year-old lions had unusually low fecundity, this might not be anything to do with being aged 3 years, it could be that food was particularly scarce 3 years previously, when these lions were cubs, resulting in that age cohort being generally of poor quality (this is a cohort effect). We can test for this cohort effect: if we repeated the study the next year, we would expect to find that 4-year-olds had low fecundity if we have a cohort effect. Alternatively, if the second study shows the same as the first, then this suggests that the low fecundity of 3-year-olds is a more general developmental effect. Hence we see that a combination of cross-sectional and longitudinal studies can be effective design technique.

If you are looking to detect developmental changes, then be on your guard for cohort effects.

3.4 Randomization

Randomization simply means drawing random samples for study from the wider population of all the possible individuals that could be in your sample.

An easy way to avoid many sources of pseudoreplication is to make sure that your experiment is properly randomized. **Randomization** is one of the simplest techniques to use in experimental biology, but is also one of the most misunderstood and abused. *Proper randomization means that any individual experimental subject has the same chance as any other individual of finding itself in each experimental group*. Properly used randomization can avoid many sources of pseudoreplication.

3.4.1 Why you often want a random sample

Let's think a bit further about why we might want to randomize. Imagine that we are interested in the effect of a genetic modification on the growth rate of tomato plants. We have fifty tomato plants that have been modified with respect to the gene we are interested in, and fifty unmodified plants. We place each plant into an individual pot of standard growing compost, and then place them all into a growth chamber. After 2 months, we harvest the plants and weigh their dry mass. There are many factors in this experimental set-up that may affect the growth rate of plants that are nothing to do with the presence or absence of the genetic modification. Maybe the growth chamber is slightly warmer or lighter at one end than the other, or maybe the quality of the compost varies between pots. Now imagine that we do not do any randomization, but instead place all the modified plants at one end of the growth chamber and all the unmodified plants at the other. Now the types of plant differ not just because of the modification, but also due to any of the number of small differences in microclimate that occur between the ends of the chamber. The measurements have become pseudoreplicates, because the effects of the treatment are confounded with the systematic differences in position within the growth chamber. This might cause us to see differences in growth rate between the two groups of plants, even if the gene modification had no effect, or might hide differences if the gene does have an effect.

Similarly, imagine that potting proceeds by taking a handful of compost from the bag, filling the pot, and then putting the plant in. Further, imagine that we start by potting all the unmodified plants first. If there are any differences in the quality of the compost as we go down the bag, this will lead to a systematic difference between the quality of the compost that the two groups are grown in. In this case, the effects of treatment are confounded with difference in compost; again, this might lead to us thinking that we find apparent differences due to genetic modification that are actually due to a difference in compost quality. Such problems can be avoided if the experiments are *properly* randomized, so that any plant has an equal probability of being put in a particular pot of compost, or being in any part of the incubator. We emphasize the word properly here, because inadequate randomization is probably the most common flaw in experiments right from undergraduate projects to those of the most eminent professors.

Careful randomization is needed in order to avoid unintentionally introducing confounding factors.

3.4.2 Haphazard sampling

The major problem that arises is that, for many people, when they say that they took a random sample, what they actually mean is that they took a haphazard sample. So what is the difference? Imagine that we have a tank full of forty hermit crabs that we wish to use in a behavioural experiment. The experiment requires that we allocate them to one of four different treatment groups. Random allocation would involve something like the following.

- Each crab would be given a number from 1 to 40.
- Pieces of paper with numbers 1 to 40 are then placed in a hat.
- Ten numbers are drawn blindly, and the crabs with these numbers allocated to treatment A.
- Ten more are drawn and allocated to treatment B, and so on until all crabs have been allocated.

Of course, we could equally have drawn the first number and put that crab in treatment A, the second in B, third in C, fourth in D, and then repeated until all treatments were filled; the details don't matter. However, what is important is that each crab has the same chance as any other of ending up in any treatment group and so all of the random variation between crabs is spread across treatments. Another way to think about a random sample is that the treatment group selected for one individual has no effect whatsoever on the treatment group selected for the next individual.

This randomization procedure contrasts with a haphazard sampling procedure. A typical haphazard sampling procedure would involve placing one's hand in the tank and grabbing a crab without consciously aiming for a particular individual. *This will not give you a random sample.* Even if you shut your eyes and think about your bank balance, this is still not going to give you a random sample. The reason for this is that there are a large number of reasons that could cause the first crabs to be picked out to be systematically different from the last crabs. Perhaps smaller crabs are better at avoiding your grasp than larger ones. 'Hold on,' you say, 'what about if I pick the crabs out in this way and then allocate them to a group without thinking; surely this will give me random groups?' Well, maybe it will, but probably it won't, depending on how good you are at not thinking. It is very tempting to subconsciously think, 'I've just allocated a crab on treatment A, so I guess the next one should be given a different treatment'. This is not random. So if you really want random groups, the only way to get them is to randomize properly, by pulling numbers from a hat, or generating random sequences on a computer. It may seem like a nuisance, and it may take you an extra half an hour, but an extra half-hour is a small price to pay for being confident that the results that you obtain after weeks of work actually mean something.

3.4.3 Self-selection

Self-selection is a real limitation to the usefulness of the phone polls beloved of newspapers and TV stations. Generally, people are encouraged to phone one number if they, for example, prefer say Coke to Pepsi, and another number if they prefer Pepsi. Now, what can you conclude if the Pepsi number receives twice as many calls as the Coke number? First you must think about the population that is sampled. The population that sees the newspaper article with the phone number is not a random cross-section of the population. People do not select newspapers in a newsagent at random, so immediately you are dealing with a non-random sample of the population. But things get worse. People who read the newspaper do not all respond by picking up the phone; indeed, only a small fraction of readers will do so. Can we be sure that these are a random sample of readers? In general, this would be very risky; generally only people with a very strong opinion will phone. So we can see that these phone polls are very poor indicators of what the wider public believe, even before you ask questions about whether the question was asked fairly or whether they have safeguards against people voting several times. Our advice is to be very careful when interpreting data that anyone collects this way, and, if you are sampling yourself, make sure that you are aware of the consequences of this sort of self-selection.

Q 3.6 What do we mean by asking the question 'fairly'? Can you give examples of unfair ways to ask the question?

In human studies, a certain amount of self-selection is almost inevitable. If you decide to ask random people in the street whether they prefer Coke or Pepsi, then people have the right to decline to answer (see section 6.7); hence they self-select themselves into or out of your study. Your experimental design should seek to minimize the extent of self-selection, then you must use your judgement to evaluate the extent to which the self-selection that remains is likely to affect your conclusions.

Self-selection is not confined to human studies. For example, if we are seeking to measure the sex ratio of a wild rodent population, then it would seem logical to set out live traps and record the sex of trapped individuals before releasing them. However, this will give an estimate for that proportion of the rodent population that are caught in traps, not necessarily the general population. It could be that some sections of the population are less attracted to the bait used in the traps than others, or less likely to be willing to enter the confined space of the trap than others. Hence the estimate of sex ratio that we get is affected by the fact that individuals may to some extent self-select themselves into our study by being more likely to be trapped.

Q 3.7 What can you do to explore how serious self-selection is for your estimate of sex ratio in the rodent population?

Q 3.8 Should we worry about pseudoreplication in the rodent trapping study described above?

Sometimes some self-selection is unavoidable, but try to minimize it, and then evaluate the consequences of any self-selection that you have left.

3.4.4 Some pitfalls associated with randomization procedures

In principle, randomization is simple, but it is important to give some thought to the exact procedure that is used, otherwise we run into some unexpected problems. With the hermit crabs, it is clear that the 'numbers in a hat' procedure that we used would ensure that each crab had the same chance as any other of ending up in any particular treatment. Put another way, for each treatment we have a group of crabs that is representative of the population as a whole. However, imagine that we wanted to do a study on territorial behaviour in great tits. In our study area there are fifty great tit territories, and we want to select ten of them for our study. We want our sample to be representative of all the territories, so we decide to use a randomization procedure to choose the territories, to avoid any possible biasing. Fortunately, we have a map of our study area with all the territories marked, which is also marked with a grid. We use the following procedure.

- We get a computer program to give us pairs of random numbers.
- From each pair of numbers, we get a randomly chosen grid reference.
- For each grid reference, we pick the territory that contains or is closest to that grid point.

However, although we have tried to be careful, this procedure will not give us a random sample. The territories are likely to differ in size, and the bigger the territory, the more grid points it covers. Thus our procedure will lead us to select territories for our sample that are on average bigger than typical territories. We would have been far better to number each of the territories, and then use a random number generator to pick ten territories.

Care must be taken to ensure that the random sample that we take is representative of the population that we want to sample.

3.4.5 Randomizing the order in which you treat replicates

The need to randomize doesn't just apply to the setting up of an experiment, but can apply equally to measuring. There are numerous reasons that can lead to the accuracy of measurements differing through time. Maybe the spectrophotometer that you are using is old, and gets less accurate as time goes on. Or maybe you have spent 10 hours staring down a microscope counting parasites, and the inevitable tiredness means that the later counts are less accurate. Or maybe after watching 50 hours of great tit courtship behaviour on video you become better at observing than you were at the beginning. Whatever the reason, this means that if you do all the measurements on one treatment group first and then all those of another treatment group, you risk introducing systematic differences between the groups because of changes in the accuracy of the methods. It is far better to organize your sampling procedure so that individuals are measured in a random order (see Chapter 5 for more about this sort of problem).

Avoid introducing time of measurement as a confounding factor.

3.4.6 Random samples and representative samples

Imagine you ask a computer program to randomly pick five unique random numbers between 1 and 100. It is possible that it will randomly generate something that looks non-random, say the set of numbers {1, 2, 3, 4, and 5}. This is very unlikely, but will happen occasionally. What should you do? If you think that this will affect the quality of your results, then you should discard this set of random numbers and ask the computer to try again. For example, imagine that over the years you have written 100 reports, and you want to measure the occurrence of the word 'however' in these. Taking a random sample of the 100 reports seems a good idea, but if that sample is numbers 1 to 5, then you are only sampling reports that you wrote very early in your career. This seems less good; hence you should ask the computer to try again. This is not cheating. One way to think about this is that your aim is not to get a random sample, but a representative sample. Almost all random samples are representative, but some are not. But be careful that you don't cheat. If you carry out statistical analyses and get a result that you find inconvenient, and you then conclude that this was because your random sample was unrepresentative, then we would be very uncomfortable about you discarding this experiment and trying again with another random sample. Ask yourself: would you also have decided that the sample was unrepresentative if the results of the statistical analyses were different? If the answer is 'no', then you are rejecting inconvenient results on spurious grounds by deciding that the sample was unrepresentative. Decide whether or not to reject a random sample immediately, generally before you have looked at the data from that sample. This stops any suggestion of cheating. Let us return to the case of the 100 reports. If the sample had been numbers 20, 40, 60, 80, and 100, this also looks odd, but is it unrepresentative? Probably not, as long as there is no regularity in the sort of reports you write. However, if every fifth report is a summary of the preceding four, then you do have a problem, as you are only sampling summary reports and not the original sources, which make up 80% of the reports you write. In this case 20, 40, 60, 80, and 100 would be an unrepresentative sample.

Ideally, your decision to reject random samples as unrepresentative should be based on a well-defined rule that was decided in advance. However, unrepresentative random samples occur so infrequently that very few scientists take the time to think up such rules that will very rarely come into operation. You can be certain that the Devil's advocate will take a very strong interest in any time you decide that a random sample is unrepresentative, so when you do discard a sample, make sure that you have a very strong case for justifying this decision.

You should discard random samples as unrepresentative only very infrequently and after very careful thought.

3.5 Selecting the appropriate number of replicates

Replication is the basis of all experimental design, and a natural question that arises in any study is: how many replicates do we need? As we saw earlier, the more replicates we have, the more confident we can be that differences between groups are real and not simply due to chance effects. So, all things being equal, we want as many replicates as possible. However, as with all things in life, all things are not equal. Increasing replication incurs costs. These costs might be financial: if an experiment involves expensive chemical reagents then doubling the number of replicates will result in a large increase in cost. More likely (and more importantly for many people), experiments will involve time costs.

Probably most important of all, if experiments involve using humans or animals (or materials obtained from humans or animals), then there may well be welfare or conservation costs of increasing sample sizes.

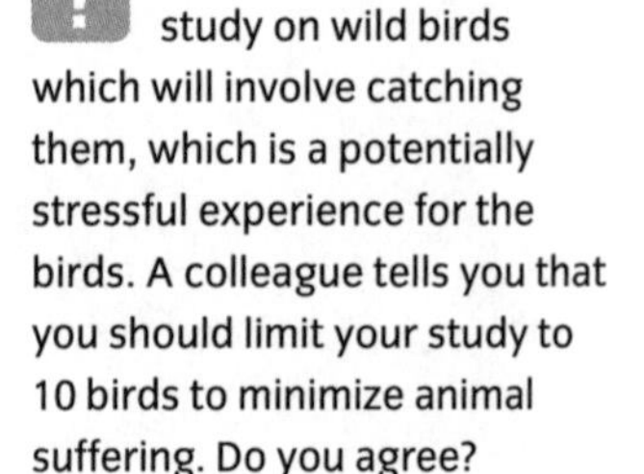

These costs mean that the general answer to the question 'How big should my experiment be?' is that it should be big enough to give you confidence that you will be able to detect any biologically meaningful effects that exist, but not so big that some sampling was unnecessary. How can we decide how big this actually is? There are two general approaches that can be used to answer this question: we can either make educated guesses based on similar studies or carry out formal power analysis.

'The more samples, the better' is an over-simplification.

3.5.1 Educated guesswork

The simplest way to get an idea of how big your experiment should be is to look at similar experiments that have been done by other people. This may seem obvious, but it is an extremely useful method for getting a ballpark figure for how big an experiment might need to be. It is also an extremely useful step in obtaining some of the biological data that you will need if you plan to go on and do a more formal power analysis. Never convince yourself that no similar experiments have ever been done before; there are *always* experiments out there similar enough to your own to be useful. This approach is simple, but very effective. We have only devoted one paragraph to it because it's a straightforward technique with no hidden pitfalls, not because we don't think it's important. We do think it is important: *go to the library*!

Learn all you can from previous studies.

3.5.2 Formal power analysis

Statistical power is an extremely simple and useful concept that can be used to guide us when we are trying to decide how big an experiment needs to be. So what is it? Statistical power is the probability that a particular experiment will detect a difference, assuming that there really is a difference to be detected. If this definition is a little terse for you, then try the example in Box 3.2; otherwise let's move on.

BOX 3.2 Returning to beetle larvae in beans to help think about power

Let's go back to the effects of competition within a bean on the size of *Callosobruchus* beetles. We are going to assume that competition really does reduce a beetle's size by an average of 1 mg. Our investigators do not know this and are engaged in their usual quest to determine whether competition has an effect on adult size, and have measured five beetles that experienced competition and five that did not. The investigators carry out an appropriate statistical test to examine whether the two groups are likely to differ. In statistical jargon, the investigators want to know whether these groups are significantly different from one another. If the investigators find a significant difference, then they conclude that competition affects size; if they find no significant difference, they conclude that there is no evidence that competition has an effect on size.

Let us imagine that the investigators get a statistically significant result, and come to the conclusion that competition has an effect on size. However, the

BOX 3.2 (***Continued***)

researchers do not stop there. Instead, being very thorough, they come back the next day and repeat the experiment in exactly the same way as they did the day before. The sizes of the individual beetles that they measure will differ from those of the day before, because of the effects of random variation (the beetles will differ in any of the huge number of other factors that affect size). This in turn will affect the mean size of the two groups of beetles. Again, the investigators do their statistics and find a significant difference between the groups. However, even this is not enough for our intrepid researchers, and they are back in the lab the following day doing exactly the same experiment again. Again, the exact size of the beetles measured differs, and this affects the means. However, today (by chance) the five beetles raised in competition all happen to have positive deviations due to random variation, while those raised without competition are all smaller than would be expected, having negative deviations. This means that there is very little difference between the means of the two sample groups, and when the investigators do the statistics there is no significant difference between the groups, so today they find no evidence that competition has an effect.

This is a strange set of circumstances, so let's get this clear. The investigators have done nothing wrong; the sample groups were all randomly chosen, and the experiment was carried out correctly. The fact that the beetles in the competition group all have positive deviations is just down to chance, like tossing a coin five times and getting five heads. It's unlikely to happen, but is always a possibility, and that means it will happen sometimes. When it does, it will lead our researchers to draw the wrong conclusions and say that there is no effect of competition when in fact there is.

However, at this point our investigators become very worried, and they come back every day for 100 days and repeat the experiment (we don't recommend you try this yourself!). On 72 days they find a statistically significant difference between samples, and on the other 28 days they find no evidence of a difference. Now we know that there really is a difference between the two populations from which all the samples were drawn. So what this means is that with their particular experimental design, the experimenters have got the correct answer on 72 out of 100 tries.

Thus the power of the experiment (the probability that it detects the difference between the populations, assuming that there is one to detect) is about 72%.

3.5.3 Factors affecting the power of an experiment

Random variation is the variation between sample units that cannot be accounted for by the factors considered and so is due to other (random) factors.

The power of an experiment of a specific design will be affected by three main things: the **effect size**, the amount of **random variation** and the number of replicates.

In the example in Box 3.2, the effect size is the real difference between the mean weights of the two populations of beetles raised with and without competition. In other cases the effect size might be the strength of the relationship between two variables. In fact, the effect size can be anything that we are trying to detect. The important thing about the effect size is that, all other things being equal, the bigger the effect size, the easier it will be to detect, and so the more powerful the experiment will be. The reason for this is obvious: if competition affects beetle weight by 1 mg, then it will be easier for this effect to be masked by random variation than if competition affects beetle weight by 3 mg.

The **effect size** is the magnitude of the effect of one factor on the variable that we are measuring. If we are looking at the factors that influence a dog's weight, then both breed and the amount of daily exercise might be expected to influence weight, but we would expect breed to have a stronger effect. If this is true, then the effect size associated with breed is higher than that associated with exercise.

The **effect size** is the magnitude of the effect that we are measuring.

An increase in the amount of random variation has the opposite effect to an increase in effect size. As the amount of random variation increases, it becomes harder to detect an effect of given size. In the beetle example, we can think of the size of a beetle as being partly due to the treatment group that it is in, and partly due to other factors. As the importance of the other factors increases, the effect of the treatment becomes harder to detect.

Finally, the size of the samples will affect the power, with more replicates increasing the power of an experiment. Again, the reasons for this will be obvious after our discussion above about the effect of replication. Replication acts to cancel out some of the effects of random variation, because with several measurements the chance of them all suffering similarly from random variation is very small.

3.5.4 Relationship between power and type I and type II errors

So statistical power depends in part on what the world is really like (i.e. how big the biological effects are and how variable the world is), and in part on how we do the experiment (i.e. how big our experiment is, and the particular design that we use). We can do very little about the first part (although we may be able to reduce the amount of random variation by doing experiments in controlled environments), but we do have control over the second part. To determine the power of our experiment, we need to make an educated guess as to what the world is like, and then we can determine the power of different types of experiments. See Box 3.3 for another thought experiment that should make this clearer.

Exactly the same procedure as in the chicken feed example in Box 3.3 can be used to determine the power of any experiment that we can imagine. In principle, all we need to do is decide what we think the world is like, and we can then simulate experiments with different sample sizes, or even different experimental designs. In practice, we generally don't have to go

BOX 3.3 An example to demonstrate the effect of experimental design on statistical power

Researchers working for an animal feed company are interested in deciding whether adding a particular type of food supplement to chicken feed will increase the weight of chickens. They decide to do a power analysis before the experiment, to estimate the number of replicates that they should use. First they must decide on the effect size. In this case, this is the amount by which the supplement will increase the average weight of a chicken. Now obviously they do not know for sure what this is (that's why they are doing the experiment), so they have to make an educated guess. Now this guess might be based on results for other similar supplements; maybe other supplements tend to increase weight by around 500 g, so the investigators expect a similar effect from their supplement. The experimenters might have carried out a small pilot study to get some idea of sizes of effects before embarking on the full experiment. Alternatively, the guess might be based on theoretical predictions. Maybe there are physiological models that suggest that this type of supplement would be expected to increase weight by 500 g. Finally, maybe there are reasons why the investigators would not be interested in an effect of below a certain size. For example, maybe a supplement that increased weight by less than 500 g would have no use in the poultry industry, so the investigators are only concerned about being able to detect an effect of at least 500 g. Whatever the reason, the investigators decide on 500 g as the effect size that they want to be able to detect in their experiment.

The next step is to estimate the amount of random variation that will be experienced. Again there are numerous ways of doing this, but the easiest is to use biological knowledge of the system (or other similar systems). Our researchers know that the body weight of chickens in their facility tends to be normally distributed, with a standard deviation of y g. The mean size of an unsupplemented chicken is x g. With these two pieces of information, the researchers have a picture of what the world might look like: chickens fed a standard diet will have weights with a mean of x g and a standard deviation of y g, while chickens fed the supplement will have a mean of $x + 500$ g and a standard deviation of y g. This picture is shown in Figure 3.2. The question for the researchers is: if the world is really like their best guess, how many chickens will they need to use in their experiment to reliably test this?

They decide that their experiment will involve two equally sized groups of chickens, one group fed the normal diet and one fed the normal diet plus the supplement. What will the power of such an experiment with five chickens in each group be? To get an idea of this, they run an imaginary experiment on a computer. To do this they get the computer to pick five chickens from a normal distribution with a mean of x g and a standard deviation of y g, and five from a similar distribution with a mean of $x + 500$ g. They then assess whether they could detect this difference in their imaginary experiment by putting these weights into a computer package, doing a statistical test and seeing if they get a significant difference between the means of the two groups. They repeat this sampling and testing procedure 1000 times (this takes virtually no time on a computer) and see how many times their imaginary experiment detects the difference. They find that in about 50% of these runs they detect the difference

BOX 3.3 (***Continued***)

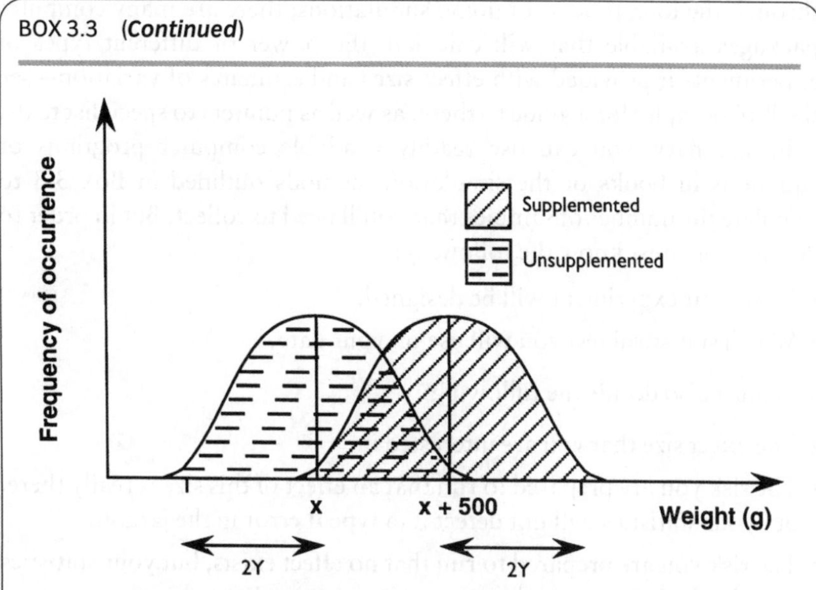

Figure 3.2 Hypothesized effect of a food supplement on the distribution of chicken weights.

and in 50% they don't, suggesting that this experiment would have a power of about 50%. In practical terms, this means that if their view of the world is correct, and they carried out the experiment with five chickens in each group, they would only have a 50 : 50 chance of detecting any difference caused by the supplement. These are not great odds, so they decide to see what would happen if they increase the number of chickens. To do this they simply repeat the previous procedure, but this time they get the computer to generate groups of 10 chickens. This time they get a power of 60%. They continue to do the same, but with samples of 15, 20, 25, and 30 chickens. They find that a sample of 30 chickens gives then a power of 87%, and so decide that this is the size of experiment that they will use. This means that *if the world is as they think it is*, and they carry out the experiment, they have an 87% chance of detecting a difference of at least 500 g between the two experimental groups (assuming that a difference actually exists). If they do the experiment and discover no difference, they can be confident that this is probably because the feed has no effect (or at least its effect is smaller than 500 g).

Of course, the investigators might decide that 500 g is a bit big, and that to be on the safe side, they will reduce the effect size to 250 g to see what effect that would have. They could use exactly the same procedure to determine the experiment's power in this alternative situation. Alternatively, they might decide that they can reduce the amount of random variation in the experiment (maybe by using genetically identical chickens, or by increasing the control of the temperature in the experiments). To see the effect that this change would have on the power of the experiment, they would simply get the computer to use the new (lower) standard deviation in its simulated experiments. Thus, by simulating experiments in imaginary worlds, the investigators are able to get a good idea about what sort of sample size they will need for their real experiment.

through the long process of doing simulations; there are many computer packages available that will calculate the power of different types of experiments if provided with effect sizes and estimates of variation—see the Bibliography for a guide to these, as well as pointers to specialist texts.

In summary, you can use readily available computer programs or equations in books or the simulation methods outlined in Box 3.3 to calculate the number of samples that you'll need to collect. But in order to do this, you must know the following:

- How your experiment will be designed.
- Which statistical test you will use on your data.

You must also decide the following:

- The effect size that you are interested in.
- The risk you are prepared to run that an effect of this size is really there, but your statistics will not detect it (a **type II error** in the jargon).
- The risk you are prepared to run that no effect exists, but your statistics mistakenly suggest that there is an effect (a **type I error**).

A **type II error** occurs if there really is an effect of one factor on the variable of interest, but your experiment fails to detect it. A **type I error** occurs if there is in fact no effect of a factor, but your experimental results (just by chance) suggest that there is.

Note that you must accept some risk of these errors occurring; there is no certainty in statistics, just acceptably low levels of risk that you've got it wrong. The probabilities of making an error of each type are linked, and to some extent under the control of the investigator. Box 3.4 looks at these two types of errors in more detail.

BOX 3.4 **Type I and Type II errors**

Let's think a little more about the different outcomes that we might get from an experiment that examines whether the size of a beetle is affected by competition. The four possible outcomes of such an experiment are shown in the table below.

What the real world is like	What our experiment detects	
	No effect of competition	Effect of competition
Competition affects size	Type II error	Correct conclusion
Competition doesn't affect size	Correct conclusion	Type I error

Let's begin by thinking about the first row in the table. Here we are assuming that competition really does affect the size of beetles (i.e. we are in the Small Bean World of Box 3.1). If our experiment detects this difference as being statistically significant (that is, unlikely to have occurred by chance), then we draw the correct conclusion that size is affected by competition. However, if we do not detect a statistically significant difference in the size of the beetles, our experiment will

BOX 3.4 (*Continued*)

lead us to draw the incorrect conclusion that competition does not affect size when it actually does. This is referred to as a type II error. We can then define the type II error rate as the probability of making a type II error, or the probability that our experiment doesn't detect a difference, when in fact there is a difference to detect. Observant readers will spot that this is the opposite of the definition of power given. In fact, power can be thought of as the probability of not making a type II error, and the two are related by the simple equation:

$$\text{Type II error rate} = (1 - \text{power})$$

Now let's consider the second row of the table, the world where competition has no effect (our so-called Big Bean World). In this situation, if our experiment does not detect any difference between the groups of beetles, we will draw the correct conclusion that competition does not affect size. However, if our experiment does detect a significant difference between the groups of beetles (because by chance we have picked unusually big beetles in one group and unusually small ones in the other), we will be mistaken in believing that competition does affect size when in fact it does not. This is referred to as a type I error, and again the type I error rate is the probability of making a type I error. The type I error rate of an experiment is entirely under the control of the experimenter, and will be determined by the significance level chosen for the statistical test. By convention, a type I error rate of 0.05 (or 5%) is regarded as acceptable in the life sciences. This means that there is a 1 in 20 chance that we will make a type I error. Now 1 in 20 might sound quite high to you—why don't we set a smaller type I error rate of 1 in 100, or 1 in 10 000? Surely this would mean that we will be wrong less often? The problem is that the type II error rate of an experiment is also in part affected by the type I error rate chosen by the experimenter. If we reduce the probability of making a type I error, we automatically increase the chance of making a type II error; we cannot simultaneously minimize both kinds of error, so in the end we have to come up with some sort of compromise, and this compromise is generally a type I error rate of 0.05. Of course, in situations where the consequences of the different types of errors differ greatly, we might choose a different value. For example, in deciding whether a substance has harmful effects on human embryos, we might decide that the consequences of failing to detect a real effect (making a type II error) are much worse than mistakenly finding an effect that does not really exist (making a type I error), and increase our type I error rate accordingly. On the other hand, for deciding whether an expensive drug is an effective treatment, we might decide that mistakenly concluding that the drug is effective when it is not would be very serious, and reduce our type I error rate.

In addition, you need to be able to estimate some measure of the variability between samples.

This seems like a long list, but it isn't really—in general only the effect size and the variability should give you any significant extra work. Further, this effort in thinking and calculating must be set against the perils of conducting an experiment with too few samples or the cost (and perhaps irresponsibility) of using too many. Power analysis for simple designs can be relatively straightforward, and the rewards for mastering it are considerable.

Summary

- Whenever we carry out an experiment, we are trying to find ways to remove or reduce the effects of random variation, so that the effects that we care about can be seen more clearly.
- Replication involves making the same manipulations of and taking the same measurements on a number of different experimental subjects.
- Replication is a way of dealing with the between-individual variation due to the random variation that will be present in any life science situation.
- In the life sciences, we are almost always working with a sample of experimental subjects, drawn from a larger population. To be independent, a measurement made on one individual should not provide any useful information about the measurement of that factor on another individual in the sample.
- If your sample subjects are not independent, then you are pseudoreplicating rather than replicating.
- Unacknowledged pseudoreplication is a very serious flaw in an experiment, but can be avoided with a little care at the planning stage.
- Pseudoreplication is an issue about biology, and it requires biological insight; you cannot turn to a statistician for salvation, but must face it yourself.
- Randomization simply means drawing random samples for study from the wider population of all the possible individuals that could be in your sample.
- Properly used randomization can control for many sources of random variation.
- Randomization needs care if you are to do it properly.
- Haphazard sampling is not random sampling.
- Self-selected individuals almost never form a random sample.
- One way to decide how many replicates to use is to make an educated guess based on previous similar studies.
- The alternative is a more formalized method called power analysis. This is not tricky and there are many computer programs that can help you.
- Too few replicates can be a disaster; too many can be a crime. So you need to think carefully about sample sizes.

Different experimental designs

4

- Having mastered the concepts of randomization and replication in the preceding chapter, you can now get to grips with some simple but powerful experimental designs.
- We begin by introducing the idea of a control group, which is an essential part of many life sciences experiments (section 4.1).
- The simplest type of design is a completely randomized design, as introduced in section 4.2.
- A completely randomized design can perform poorly in cases where random variation between individuals is high. If a given variable is expected to introduce significant unwanted variation to our results, then we can control that variation by blocking on that variable. Such blocked designs are introduced in section 4.3.
- In paired designs, we divide the sample into pairs and randomly assign them, one to each of two groups. In section 4.3.3, we introduce paired designs as an extreme form of blocking.
- In a within-subject design, experimental units experience the different experimental treatments sequentially, and comparisons are made on the same individual at different times, rather than between different individuals at the same time. These designs are introduced in section 4.4.
- Split-plot designs, which can be a convenient way of dealing with experiments involving more than one experimental factor, are introduced in section 4.5.
- The design that you choose and the statistics that you will use are closely linked, and so thinking about both in advance is essential (section 4.6).

4.1 Controls

Suppose that we claim that buying this book will improve your final degree mark. You decide that anything that will improve your mark has to be worth such a modest investment, and so buy the book (and maybe even read it). Subsequently, when the exam results are announced, you are delighted to find that you have done better than you expected. Does this mean that our claim was true? Well, you may feel that the book helped you to design better experiments or improved your ability to critically read the literature (or helped you get to sleep at night). Indeed, we hope there are many reasons why this book will help to improve your scientific abilities. However, based only on your experience it is hard to say that the book had any effect on your degree mark at all. The reason is that we don't know what degree mark you would have got if you hadn't bought the book. Your experiment to examine the effect of buying this book on your exam marks is flawed because it has no **control**.

A **control** is a reference against which the results of an experimental manipulation can be compared. For example, if we wish to explore the effect of smoking on human lung tissue, we need a **control group** of tissue samples from non-smokers to compare with our **treatment group** of tissue samples from smokers.

4.1.1 Different types of control

Most, but not all (see section 4.1.5), experiments require a control. If you have followed our advice from Chapter 2 by clearly stating the specific hypothesis under test, then identifying whether a control is necessary, and what that control should be, is relatively easy. Consider the following hypothesis:

> Giving a vitamin supplement with the food of caged rats leads to increased longevity.

In order to test this, we need to measure the longevity of rats to which we give the vitamin supplement. However, this alone would not be enough to answer the question, because we don't know how long the rats would have lived if they had not been given the supplement. We need a control. In an ideal world, we would compare the longevity of the same rats both when they have been given a supplement and when they have not. However, a rat can only die once, so clearly this is not an option. Instead we need a **control group** of rats that are identical in every way to our **treatment group** (sometimes called experimental group), except that they do not experience the experimental manipulation itself. By 'identical' we mean in every practical way: the rats must be of the same strain, from the same source, kept under the same conditions. This is easily done by randomly assigning individuals from a common pool to the two groups: treatment and control (see Chapter 3). The control group allows us to estimate the effect of the supplement, because we can see what happens without the supplement.

Very often, a control group is a group to which no manipulation is applied. This is called a **negative control**. However, sometimes, the research question will call for the control group to also be manipulated in some way; this is called a **positive control**.

The rats example above used a **negative control.** To see the potential use of a **positive control**, consider the hypothesis:

> A particular novel treatment for warts in humans produces better results that the currently used method.

In this case, we must measure the clear-up rates of warts in a group of patients given the novel treatment. However, to compare these rates with those of patients given no treatment at all would not test the hypothesis posed; rather, it would address the hypothesis:

> The novel method produces a beneficial effect on the clear-up of warts, compared to taking no action.

To address the original hypothesis, we need a control group of patients that are given the established method of treatment, rather than no treatment at all.

Of course there is no reason why an experiment cannot have both types of control. If we just have the positive control of treating people with the new treatment or the established treatment, we can say which treatment is better. It is unlikely, but possible, that both treatments actually make warts worse, and one is just less bad than the other. The inclusion of a third, negative, control group would allow this possibility to be examined, and (if the results turned out that way) for us to say with confidence that the new method was not only less bad at removing warts than the existing treatment, but that both were actually better than not doing anything at all.

Is a control really necessary in this example? You might argue that there are probably already studies on the effectiveness of the existing treatment. Could we not forget about using our own control and simply compare the results of our treatment group of patients with the published results of previous studies of the effectiveness of the established method? In the jargon, this would be using a **historical control** rather than a **concurrent control**. This approach has the attractions of requiring less effort and expense, two worthwhile aims in any study. However, there are potentially very serious drawbacks. The key thing about the control group must be that it differs from the treatment group in no way except for the treatment being tested (this is to avoid confounding factors and problems of pseudoreplication; see sections 1.4 and 3.3). This is easier to achieve with a concurrent control than a historical one. Imagine that we found that the new treatment cleared up people's warts in an average of 2 weeks, while a study carried out 3 years ago in another hospital found that the original treatment cleared warts in 4 weeks. It might be that the difference between these two studies is due to the differences in treatment. However, it could also be due to anything else that differed systematically between the studies. Maybe the individuals in one study had more serious warts, or a different type of wart. What if the techniques of applying the treatments differed, or the definitions of when warts had cleared up differed? All of these other confounding variables mean that we can have little confidence that the apparent differences between the two trials are really due to the different treatments. We may have saved effort and expense by using a historical control, but with little confidence in our results, this is clearly a false economy. A historical control can certainly save time and resources,

If we run our own control group, this is called a (positive or negative) **concurrent control**; if instead we use historical data as a reference, this is called a **historical control**.

Q 4.1 One of us once heard a colleague say that they were pleased to have finished their experimental work, and that all they had left to do was run their control groups. Should they be worried?

and may have ethical attractions, but have a long think about how valid a control it is for your study before deciding to adopt a historical rather than a concurrent control.

Careful statement of the hypothesis under test makes it easy to determine what type of control your experiment requires.

4.1.2 Blind procedures

In our study of the effects of the different wart treatments—let us imagine that they are both medicated creams—we need to make some assessment of when the warts have been cleared up. Now it would obviously make no sense if we use different criteria for the two different creams. However, what if the person that is making the assessment has some preconceived feeling about whether the new cream was likely to be effective or not. If this person also knows which people got which cream, then their prejudice may bias their assessment of when the warts are gone from a patient. Sometimes this might be deliberate deception on behalf of the researcher, although this is not usually the case. However, even if the researcher is not being deliberately dishonest, there is still the possibility that they will unconsciously bias the assessment in line with their expectations. The obvious solution is to organize your experiment so that the person who judges the improvement in the patient as a result of treatment does not know which of the two treatments a particular patient received, or even that there actually are two different groups of patients. This is called a **blind procedure**, since the assessor is blind to the treatment group that an individual belongs to. The attraction of blind procedures is that they remove concern that the assessor may consciously or subconsciously bias their assessment.

A **blind procedure** is one in which the person measuring experimental subjects has no knowledge of which experimental manipulation each subject has experienced or which treatment group they belong to. In experiments with humans, we may use a **double-blind procedure** in which the experimental subjects too are kept ignorant of which treatment group they belong to.

In a blind study such as this, it is critical that, although the assessor is 'blind', someone knows which treatment group the patients belong to, or else the data collected wouldn't be any use. Such restricted access to information is often achieved by using codes. For example, each patient in the example above can be given an individual code number that can be entered on their patient record and accessed without restriction, provided that someone (who should preferably not be involved with the assigning of patients to treatment groups, the treatment of the patients or their final assessment) has a list that identifies which treatment group an individual code number refers to.

You should not worry that your collaborators will think that you are accusing them of being fraudsters if you suggest that blind procedures be used. These procedures are designed to remove the *perception* that *unconscious* bias *might* taint the results of a study. The Devil's advocate will happily thrive on such a perception, and no-one can reasonably claim that it would be impossible for them to have an unconscious bias.

Blind procedures are a particularly good idea whenever the experimenter is required to make any subjective measurements on experimental units, although even something as apparently objective as timing with a stopwatch can be affected by observer bias (see section 5.2). When the experimental units are humans, then it may also be preferable to go one step further and ensure that the subjects themselves are blind to the treatment group that they are in. Such a procedure is called **double-blind.** To see the advantage of such a precaution, return to the wart treatments. Imagine that recent newspaper publicity has cast doubt on the effectiveness of the established method of treatment. It might be that patients assigned this treatment might be less motivated to apply the cream regularly and carefully, compared to those assigned to a new treatment that has escaped any negative publicity. This difference between the two groups could falsely inflate the apparent effectiveness of the novel treatment.

A **placebo**, or **vehicle control**, is a treatment that is designed to appear exactly like the real treatment except for the parameter under investigation.

In order to facilitate blind and double-blind experiments, human patients are sometimes given **placebos**. For example, a placebo tablet would be made that was identical to the real tablet, except that the active ingredient under study would not be added. Note that care should be taken so that the other ingredients are identical to the real tablet, and that the tablet and its packaging look the same (although a code would be used to identify it as placebo or real to the person finally analysing the experiment).

Blind procedures involve a bit of extra work, but if your experiments involve humans as subjects or measurement tools then they are vital.

4.1.3 Making sure that the control is as reliable as possible

If we want to have the maximum confidence in our experimental results, then the control manipulation should be as like the experimental manipulation as possible. Sometimes this will require some careful thought into exactly what control we need. Otherwise we might not avoid all possible confounding factors.

For example, consider an experiment to see whether incubating an artificially enlarged clutch of eggs changes the nest attentiveness of a parent bird. We will need a group of control nests in which the clutch size is not enlarged. Let us imagine that these controls are simply nests

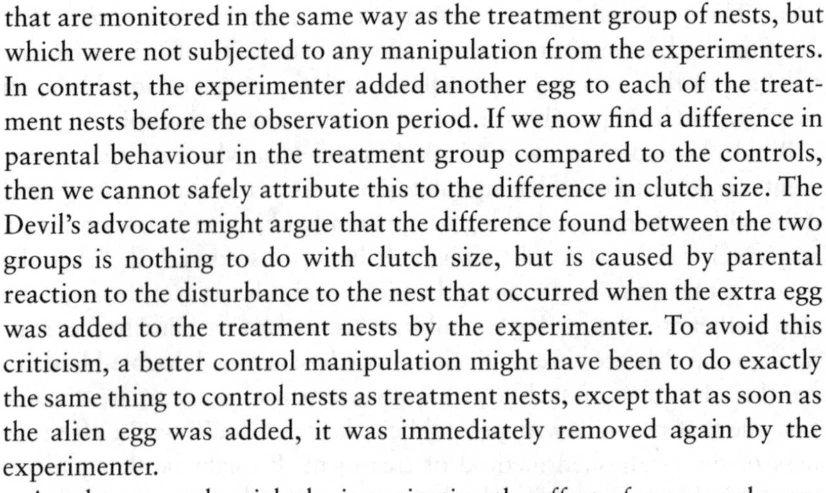

that are monitored in the same way as the treatment group of nests, but which were not subjected to any manipulation from the experimenters. In contrast, the experimenter added another egg to each of the treatment nests before the observation period. If we now find a difference in parental behaviour in the treatment group compared to the controls, then we cannot safely attribute this to the difference in clutch size. The Devil's advocate might argue that the difference found between the two groups is nothing to do with clutch size, but is caused by parental reaction to the disturbance to the nest that occurred when the extra egg was added to the treatment nests by the experimenter. To avoid this criticism, a better control manipulation might have been to do exactly the same thing to control nests as treatment nests, except that as soon as the alien egg was added, it was immediately removed again by the experimenter.

Q 4.2 Can you see any benefit to a blind procedure in the bird's nest study?

Another example might be investigating the effect of a pacemaker on the longevity of laboratory rats. Now applying a surgical procedure to a rat is likely to be stressful, and this stress may have effects on longevity that are nothing to do with the pacemaker. You might consider subjecting the control group to the same surgical procedures as the treatment group, with the sole exception that the pacemaker is not added, or is added and then immediately removed. This controls for any confounding effects due to the anaesthetic or other aspects of the surgical procedure. Although this type of control has scientific attractions, ethical considerations may count against it (see section 4.1.4).

You might feel a bit of a fool going through what seem like elaborate charades (or sham procedures as the jargon would have it). You should not; the fool is the person who does not strive for the best possible control.

One last thing: make sure that all manipulations and measurements on the treatment and control experimental units are carried out in a random order. If you measure all the manipulated units and then all the controls, then you introduce 'time of measurement' as a potential confounding factor (see section 5.1 for more on this).

Choosing the best control group for your experiment warrants careful thought.

4.1.4 The ethics of controlling

Sometimes a negative control may be unethical. In medical or veterinary work, deliberately withholding treatment to those in need would be hard to justify. In the example above involving surgical procedures on rats, ethical considerations may cause you to reject the type of control suggested in favour of one that involved less animal suffering. For example, if you could argue (based on previous work) that the surgical procedures themselves (rather than the pacemaker) were highly unlikely to affect longevity, then a control group of unmanipulated animals might be the best compromise between scientific rigour and ethical considerations. Further, if we were dealing with a species where even keeping the animals in captivity was stressful, then we might even consider using a historical control (if one were available) in order to avoid having to keep a control group. But such choices must be made *extremely* carefully. While the use of historical controls may reduce the suffering of animals in the experiment—an aim which we cannot emphasize too strongly—if the controls are inappropriate and will not convince others, then *all* of the animals in the experiment will have suffered for nothing and we would have been better not doing the experiment at all.

Be careful that you don't adopt a poor historical control for ethical reasons; better you do no experiment at all than a poor one.

4.1.5 Situations where a control is not required

In view of what we've discussed so far, you might be surprised at our suggestion that sometimes a control is not necessary. To see why this might be the case, consider answering the following research question:

> Which of three varieties of carrot grows best in particular conditions?

Here there is no need for a control, as we are only comparing growth rates between the three treatment groups; we are not comparing them with some other growth rate. Similarly, there is no need for a control in an experiment to address the following:

> How does the regularity of feeding with a liquid fertilizer affect the growth of tomato plants?

Here, we again want to compare different treatment groups (i.e. groups of plants fed at different rates), and there is no need for a separate control. However, the selection of appropriate treatment groups to answer such questions is an art in itself (see section 6.1).

Sometimes, but not often, a control group is not required. A clearly stated hypothesis will help you spot such cases.

4.2 Completely randomized and factorial experiments

Unfortunately, the experimental design literature is littered with jargon and this can make understanding experimental design seem daunting. However, there is nothing really difficult about the ideas behind the terms, so we will begin by explaining some of the commonest terms. Let us imagine for the tomato trial of the last section that we decide that five different feeding rates should be explored, and that we should feed 20 plants at each of these rates. This means that we require 100 plants in total. Remembering the importance of randomization from Chapter 3, we decide to allocate our plants at random to the five different treatment groups. One way to do this would be to take each plant in turn and get a computer to pick a random number between 1 and 5. We then allocate the plant to the treatment with that number. This procedure would certainly achieve a random allocation of plants, but because of the vagaries of random sampling we are very likely to end up with slightly different numbers of plants in each treatment group, perhaps 21 in one and 19 in another. In statistical jargon, our experiment is **unbalanced**. This is not a fatal flaw in our experiment, but it does have drawbacks, because generally the statistical methods that we will ultimately use to analyse the data are most powerful, and least influenced by violations of assumptions, when there are equal numbers in each group. Only in unusual circumstances (see section 6.3) should we deliberately seek to use unequal numbers. This aiming for equal numbers, or as a statistician would say for a **balanced design**, should be a general tenet of your experimental designs. A better way to assign the plants would be to add a constraint that the final number in each group must be 20. Practically, this could be done by numbering all of the plants, then putting these numbers on identical pieces of card in a bag, mixing them thoroughly and then drawing them out. The first 20 numbers to be drawn would be assigned to the first treatment group, and so on.

A **balanced** experimental design has equal numbers of experimental units in each treatment group; an **unbalanced** design does not.

An ***n*-factor** or ***n*-way design** varies *n* different independent factors and measures the responses to these manipulations. For example, an experiment looking at the effects of both diet and exercise regime on dogs' health would be a two-factor design. If we consider three different diets, then there are three **levels of the experimental factor** 'diet'. If *n* is greater than one (i.e. multiple factors are involved), then the experiment can be referred to as a **factorial** experiment.

It will not surprise you that such experiments are called **completely randomized designs**, since individuals are assigned to treatment groups completely at random. At the end of the experiment we look for differences between the groups of individuals. Statistically this can be done using a *t*-test if there are only two groups and ANOVA (analysis of variance) if there are more than two. This experiment is one of the simplest designs that you will come across: we have varied a single factor experimentally (the feeding rate) and then looked for differences between groups that experienced different levels of the factor (different feeding rates). Such a design is known as a **one-factor design** (you will also see it referred to as a **one-way design**). We have five different treatments—the five feeding rates—and we refer to these as five **levels** of our experimental factor.

So if we want to impress people with our statistical expertise, we can describe the experiment as a completely randomized one-factor design, with five levels of the factor. If that doesn't impress them, there is one further thing we can add. When we set up the experiment we made sure that we had several plants in each treatment group, and the same number in each as well. This means that our experiment is fully replicated and balanced. So we can describe our simple experiment as a balanced, fully replicated, one-factor design with five levels of the factor. You can see how things can become very complicated very quickly, but the underlying logic is straightforward.

Always aim to balance your experiments, unless you have a very good reason not to.

4.2.1 Experiments with several factors

Now let's make the tomato plant experiment a bit more complicated, say by introducing another type of tomato plant, one that has been genetically modified from our original strain. We repeat the set-up that we used for the first strain of plant with the second. Now we are varying two factors, the level of feeding and the type of tomato. This sort of design is called a **two-factor design** (or **two-way design**). If we have set up the experiment identically for the two strains, we will have what is known as a fully cross-factored (or **fully crossed**) design. This means that we have all possible combinations of the two factors; or to put it another way, we have individuals of both strains, at all feeding rates. In contrast, if we only applied three of the feeding rates to the new strain, but still used all five for our original strain, then our design would have been described as an **incomplete design**. The statistical analysis of fully cross-factored designs is easy to do, whereas analysing incomplete designs is substantially more difficult. Avoid incomplete designs whenever possible. If you need to use an incomplete design for practical reasons (such as a shortage of samples) then get advice from a statistician first.

A **fully crossed** design means that all possible combinations of treatments of the factors are implemented. Thus if, in a three-factor experiment, factor A has two levels, factor B has three levels and factor C has five levels, then a fully crossed design will involve 30 different treatment groups ($2 \times 3 \times 5$).

Thus we have arrived at a fully cross-factored, two-factor design, with five levels for the first factor, and two for the second. This will give us ten treatment groups in total (five multiplied by two), and, as long as we have the same number of plants in each treatment, and more than one plant in each treatment too, our design will be replicated and balanced. You can easily imagine how you can get three-factor, four-factor and even more complicated designs. In general, it is best just to imagine these; for experiments that you actually have to do, strive to keep them as simple as possible.

If simplicity is the key, then you might ask, 'Why do a two-way factorial design in the case above; why not do separate one-way experiments for the two different species?' There are two main attractions of using a two-way design. The first is that it allows us to investigate the **main effects** (whether growth rate increases with feeding rate in both species, and whether one species grows faster than another when both are given the same feeding rate) in a single experiment. Doing the two separate experiments does not let us compare species effectively (because the conditions may be different in the two experiments; remember the problems of historical controls discussed earlier). It would also probably require more plants in total to obtain the same statistical power. Secondly, the factorial design allows us to examine the **interaction effect** (whether the two types of plant differ in the effect that feeding rate has on their growth rate). In many cases, the interaction effect is actually what we are interested in, and in that case we have no choice but to use a two-factor design. Interactions are really interesting, but there are pitfalls to avoid when interpreting interactions, and so we explore this in detail in Box 4.1.

Let us imagine that we wish to investigate whether dependent factor *A* is influenced by independent factors *B* and *C*. If the value of *A* is affected by the value of *B* when the value of *C* is kept constant, then we can say that there is a **main effect** due to factor *B*. Similarly, if *A* is affected by *C* in a set of trials where we hold *B* constant, then there is a main effect due to *C*. If the effect of *B* on *A* is affected by the value of *C*, or equivalently if the effect of *C* on *A* is affected by the value of *B*, then there is an **interaction** between the factors *B* and *C*.

Avoid varying too many factors simultaneously, because such an experiment will necessarily be very large, and require a large number of experimental units and a large amount of effort. Also, interpreting interactions between more than two factors can be very challenging.

Q 4.3 Let's try to illustrate the problem of interpreting an interaction involving three factors. Imagine that in the tomato plant experiment we explore the effects of strain, food level and insecticide treatment on growth rate. Our subsequent statistical analysis informs us that there is an interaction between all three factors. Can you describe biologically what this means?

4.2.2 Confusing levels and factors

One confusion that can arise with complicated designs is the distinction between different factors and different levels of a factor. In the tomato example, it is easy to see that the different feeding levels are different levels of a single factor. In contrast, what if we had done an experiment with five different brands of plant feed? Now it is less clear: are the plant feeds different factors or different levels of the same factor? If we had done this experiment and had a treatment for each plant feed, then we must think of the five plant feed treatments as five levels of a factor called plant feed type, and we would have a one-factor design. But suppose we then carried out an experiment with only two of the plant feeds and set up the following four treatments: no plant feed, plant feed A only, plant feed B only, plant feed A and plant feed B together. Now it is better to view the plant feeds (A and B) as two separate factors with two levels (presence or absence), and analyse the experiment as a two-factor design. In Figure 4.2, we show a number of possible designs with their descriptions to help you get used to this terminology.

Take time to avoid confusing levels and factors.

BOX 4.1 Interactions and main effects

Spotting interactions

In the main text we claimed that one advantage of carrying out experiments with two or more factors was that this would allow you to look at both the **main effects** of the factors and also their **interactions**. Understanding what these actually are is fundamental to understanding these more complex designs. However, these concepts are also some of the most misunderstood of all statistical ideas. In this box we will explore these ideas in a little more detail in the hope of showing that they are both easy to understand and extremely useful.

Let's go back to thinking about the effects of feeding treatment on growth rate in different strains of tomato plants. Imagine that we want to investigate two feeding regimes, high and low food, and we want to compare two strains of tomato, which we will call strains A and B. This will give us four different treatments (both strains with both feeding regimes). Figure 4.1 shows several of the possible outcomes of an experiment like this.

Let's begin with the situation in Figure 4.1(a). Here we can see that for both strains of tomato, the high-feed treatment leads to faster growth than the low-feed treatment. However, if we compare the two strains of tomato when they are fed identically, we see no difference. Statistically, we would say that there is a significant main effect due to feed treatment, but no main effect due to tomato variety. Biologically we would say that growth rate is affected by how much you feed the tomatoes, but not by the particular tomato variety.

The situation in Figure 4.1(b) is a little more complicated. Here again, we see that for both varieties, the high-feed treatment leads to faster growth than the low-feed treatment. However, if we now compare the two varieties under the same feed treatment, we find that variety A grows more slowly than B. Statistically, we would now say that we have significant main effects due to both feed treatment and tomato variety; biologically we would say that tomato growth rate is affected by how much you feed them and also by the tomato variety.

Figure 4.1(c) looks very similar, but differs in one very important respect. Again, we can see that for both varieties, high feeding increases growth rate. Similarly, if we look at the growth rate of the two varieties when they are given lots of food, we see that, as in the previous case, variety A grows more slowly than B. However, if we compare the varieties at the low-feed level, we see no difference between them. What does this mean? This is what statisticians refer to as an interaction. Biologically it means that the effect of the variety of tomato on growth rate is different under different feeding regimes, or, to be a bit more technical, the effect of one factor (variety) depends on the level of the other factor (feeding level). In our example the difference is quite extreme; there is no difference between the strains at one food level, but a clear difference at another.

In Figure 4.1(d) we see a less extreme case. We do see a difference between the strains under both high- and low-food regimes, but under low food the difference between the strains is much smaller. Statistically, this is still an interaction: the effect of one factor still depends on the level of the other, or to put it biologically, the effect of variety on growth rates depends on which food treatment we consider.

We might even see a third type of interaction, as shown in Figure 4.1(e). Here the growth rate of variety A is lower than that of B when they are given lots of

BOX 4.1 (*Continued*)

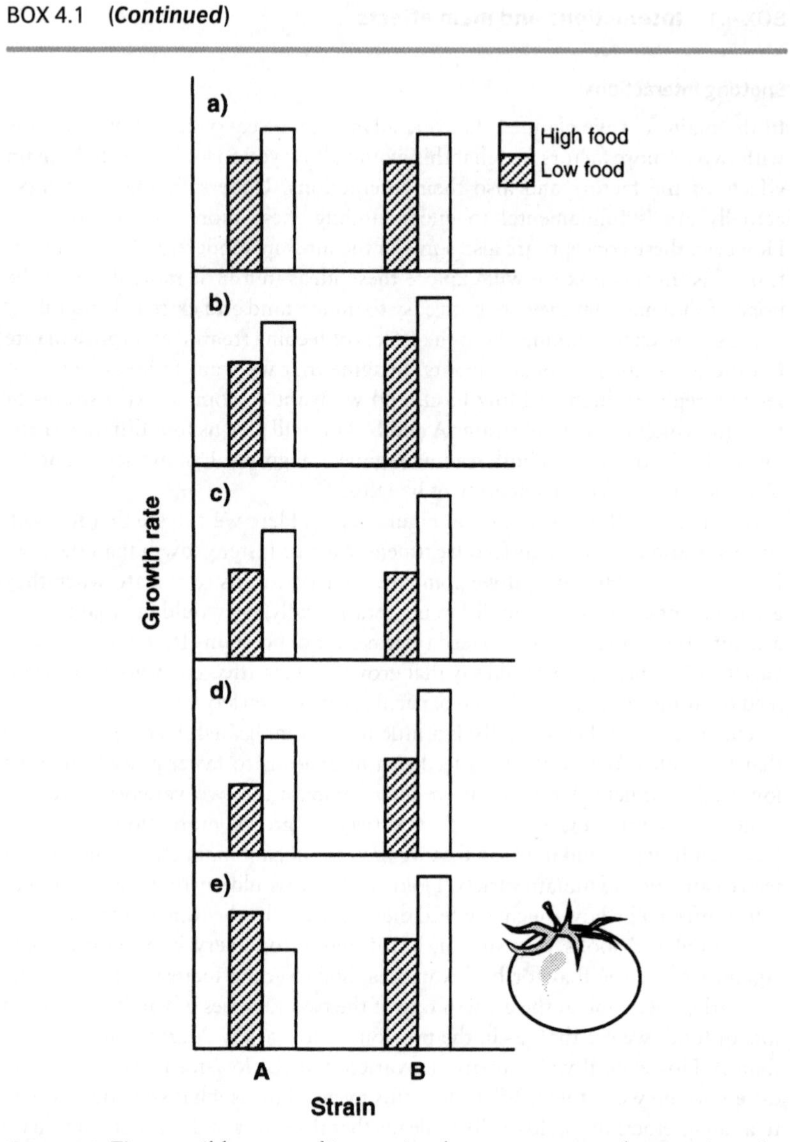

Figure 4.1 Five possible types of interaction between tomato plant strain and feeding regime.

food, but higher than that of B when they are given only a little food, so not only does the size of the difference in growth rate depend on the food treatment, but so does the direction of the effect.

So why should we care about interactions? One reason is that a significant interaction can make it very difficult to interpret (or to even talk about) the main effect of a treatment. Imagine we got the result in Figure 4.1(e). What is the main effect of food level? In one variety it is positive, and the other negative, so it is meaningless to talk about the main effect of food level. It only makes sense to

BOX 4.1 ***(Continued)***

talk about the effect of food if we state which variety we are referring to. Similarly, in Figure 4.1(d), if we want to talk about the difference between the species, we need to state which feeding regime we are comparing them under. We can't really talk about an overall difference between the species in any meaningful way. Even in Figure 4.1(b) it is hard to say exactly what the main effect of feeding is, because the size of the effect is different for the two varieties. So we can say that increased feeding has an overall positive effect on growth rate, but that the size of the positive effect depends on the variety. Only in situations like Figure 4.1(a) can we make a confident statement, such as the overall effect of increased feeding is to increase growth rate by a specified number of units. Interpretation of interactions can sometimes be tricky, and we discuss this further in section 6.6.

4.2.3 Pros and cons of complete randomization

The attraction of complete randomization is that it is very simple to design. It also has the attraction that the statistics that will eventually be used on the data (t-test or ANOVA) are simple and robust to differences in sample sizes, so having missing values is less of a problem.

Also, in contrast to some designs that we will see later, each experimental unit undergoes only one manipulation. This means that the experiment can be performed quickly and that ethical drawbacks of multiple procedures or long-term confinement are minimized.

It tends to be the case that the probability that an individual will drop out of the study increases with the length of time over which they need to be monitored, so the drop-out rate should be lower than in other types of experiments.

Imagine that you have 20 plants for a growth trial, and you split them randomly into four groups, each of which is given a different feed treatment. If all goes well, you will have 20 growth rates to analyse at the end of the experiment. However, if a plant becomes diseased, or is knocked off the bench or some other catastrophe befalls it, such that the growth rate of that plant cannot be obtained or no longer provides a fair reflection of the treatment, then you will have fewer measurements to work with. The lost or omitted cases are generally called **missing values** or **drop-out** cases. Obviously, experiments should be designed to minimize the likelihood of such drop-outs occurring. However, accidents do happen, so you should seek experimental designs where a small number of drop-outs do not have a devastating effect on the usefulness of the remaining data.

The big drawback to full randomization is that we are comparing between individuals. As we saw in the last chapter, between-individual variation due to random factors can make it difficult for us to detect the effects of the manipulations that we have carried out. If the growth rate of

1.

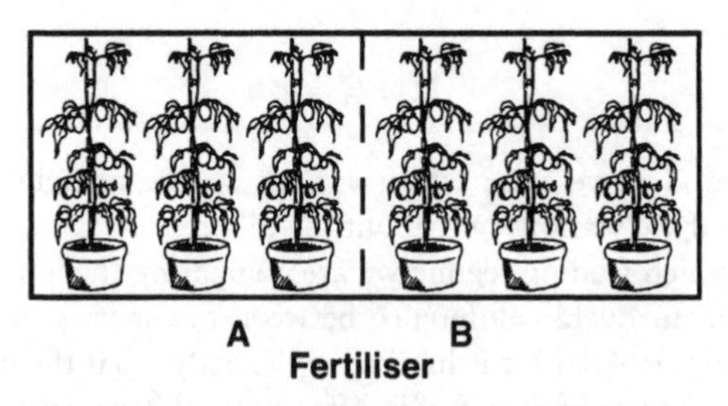

Replicated l-factor design with 2 levels of the factor (fertiliser type).

This can answer the question:

a) Do the fertilisers differ in their effect?

2.

l-factor design with 3 levels of the factor (type of cultivation).

This can answer the questions:

a) Does fertiliser affect plant growth?
b) Does pesticide affect plant growth?
c) Do fertiliser and pesticide differ in their effect on plant growth?

3.

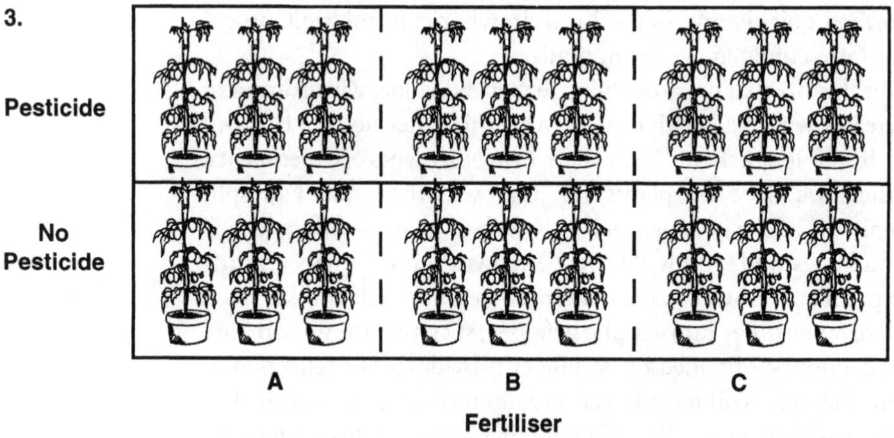

Pesticide

No Pesticide

A B C

Fertiliser

2-factor design with 3 levels of the 1st factor (fertiliser type) and 2 of the 2nd factor (pesticide use).

This can answer the questions:

a) Do the fertilisers differ in their effect on plant growth?
b) Does pesticide affect growth rate?
c) Does the effect of pesticides depend on the type of fertiliser?

Figure 4.2 Comparison of the types of questions that can be addressed by different designs.

tomato plants is highly variable due to other factors, this may make it very hard for us to detect the effects of our different feeding regimes. One solution to this problem that we have already come across is to increase sample sizes. However, this may not be the best solution to the problem. First, there is the obvious problem that this will increase the number of organisms used in our experiments, which may have ethical, conservation or financial implications (and the experiment will also use more of our time).

A second problem is that increasing the sample size can sometimes increase between-individual variation. If increasing sample sizes means that we must use experimental individuals from a wider range of sources, or that husbandry or experimental measurements must be carried out by more than one person, or if the experiment becomes so large that it requires to be housed in two places rather than one, this may increase the amount of variation between individuals. For this reason, there is an attraction towards more complicated experimental designs that reduce the amount of intrinsic variation between the experimental units being compared, or designs that require you to make comparisons between different treatments applied to the same subject individual.

If between-individual variation is low, then complete randomization can be quite powerful, hence it is commonly used in laboratory studies. It is also attractive if you expect high drop-out rates among experimental units. Field studies and clinical trials tend to suffer more from between-individual variation, and so are more likely to adopt one of the designs in the following sections.

4.3 Blocking

Suppose that we are interested in whether the type of food we give to a greyhound affects its running speed. We have at our disposal 80 greyhounds and a running track, and we want to test the effects of four different food types. We could set up a fully randomized one-factor design by allocating dogs at random to one of the four food treatments, in exactly the same way as we did with the tomato plants previously. However, this might not be the best way to set up such an experiment. What if the 80 greyhounds are of different ages, and age has a strong effect on running speed? By using a random design, we have ignored this source of variation and it has become part of the random noise in the experiment. If age has a big effect on speed, then this will mean that our experiment has lots of random noise, and we may not be able to see any effect of the different

If a given variable is expected to introduce significant variation to our results, then we can control that variation by **blocking** on that variable. This involves splitting our experimental units up into blocks, such that each block is a collection of individuals that have similar values of the blocking variable. We then take each block in turn, and for each block distribute individuals randomly between treatment groups. It is thus a more complicated alternative to complete randomization. Blocked designs are sometimes called matched-subject designs, and hence blocking can be called matching.

foods unless this effect is large. An alternative approach would be to treat age as a **blocking** factor in the experiment, as described below.

First of all, we rank the dogs by age, and then partition this ranking so as to divide the dogs into blocks, so that those in a block have a similar age. This might be 20 blocks of 4 dogs, 10 blocks of 8, or 5 blocks of 16. This choice does not matter too much (but see section 4.3.4), providing that the number of individuals in a block is a multiple of the number of groups. We then take each block in turn and randomly allocate all the dogs from one block between the treatment groups, in exactly the way that we did in the fully randomized design above. We repeat the same procedure with all of the blocks. So what does this achieve? One way to think of it is this: in the randomized experiment we will find ourselves comparing an old dog in one treatment and a young dog in another. If we compare their running speeds and find a difference, part of this may be due to food, and part due to age (and the rest due to random noise). In contrast, in the blocked design we are making comparisons within our experimental blocks, so we will only ever be comparing dogs of similar ages. With differences due to age much reduced in any given comparison, any effects due to diet will be much clearer (see Figure 4.3). The blocking has

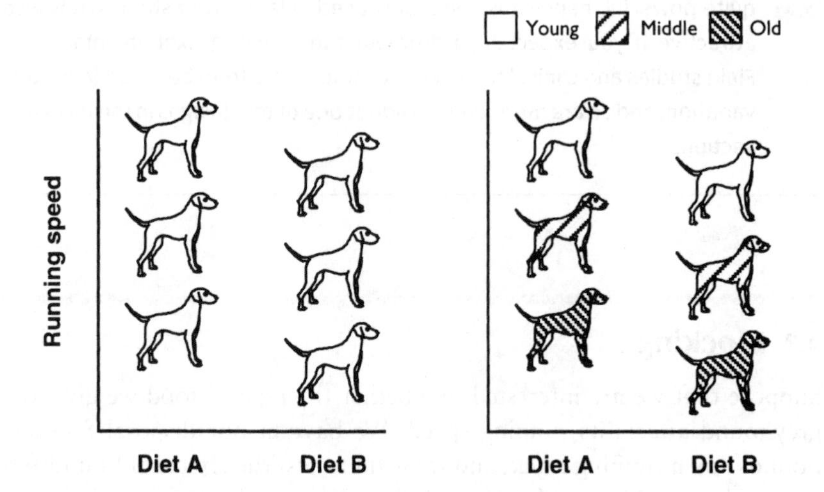

To illustrate the use of blocking, consider the following thought experiment.

In the first panel we can see the running speeds of 6 dogs, 3 on diet A and 3 on diet B. Does diet have an effect? It is hard to say.

The second panel shows the same experiment but with the dogs blocked by age: 2 old ones, 2 middle-aged ones and 2 youngsters (shown by their shading). Now the effects of diet are clearer. For all the age groups, the dog on diet A runs faster.

Figure 4.3 Schematic representation of the effect of blocking.

effectively controlled for a substantial part of the effect of age on running speed. You can think of this blocked design as a two-factor design, with food type as the first factor, and age block as the second.

Effectively what we are doing is dividing the between-individual variation into variation between blocks and variation within blocks. If we have blocked on a variable that does have an important effect on between-individual variation, then the between-block differences will be much greater than those within blocks. This is important, because in our final statistical analysis we can use 'blocks' as an additional factor in our ANOVA. This will allow us to test for between-treatment differences within each block (where between-individual differences are small) rather than across the whole sample (where between-individual differences would be larger).

If you know and can measure some factor of experimental units that is likely to explain a substantial fraction of between-subject variation, then it can be effective to block on that factor.

4.3.1 Blocking on individual characters, space and time

Remember that you can block on any variable (indeed on any number of variables) that you think might be important in contributing to variation between individuals given the same experimental treatment. The only condition for a characteristic to be used as a blocking variable is that you can measure it in individuals, so that you can rank them. We will refer to this as blocking by individual characters.

There is a second type of blocking that is commonly used in biological experiments that we will call blocking in space. Imagine that in our tomato feed experiment we find that we can't fit all our plants into a single greenhouse, but instead need to use three different greenhouses. If there are differences between the greenhouses that affect growth rates, then by carrying out the experiment at three different sites we are adding some noise due to site differences. We could simply ignore this variation and use a fully randomized design where every plant is allocated at random to a greenhouse irrespective of its treatment. However, this could be a very inefficient way to carry out such experiment. We cannot simply have all of the individuals in a given greenhouse getting the same treatment, as this would introduce greenhouse as a confounded factor. A common solution to this problem is to block by greenhouse. What this means is that we allocate equal numbers of each treatment to each greenhouse, and note down which greenhouse a plant is grown in. We now have a two-factor design with feeding rate as the first factor, and greenhouse as a second, 'blocking' factor. This means that we can make comparisons between plants in the same greenhouse, comparisons that will be unaffected by any differences between greenhouse. Blocking by site is frequently seen in large agricultural trials (where several fields or even farms may act as the blocks), but could equally apply to any experiment where assays are carried out at different sites. Sites can be anything from fish tanks, to growth shelves, to labs in different countries.

The final kind of blocking that we will discuss is blocking in time. This is exactly analogous to blocking in space, except that the assays are carried out at different times rather than in different places. This may be useful if you can't possibly assay your whole experiment in a single time period.

The recipes described above, where equal numbers of individuals from each block are assigned to all of the treatment types (or equal numbers of each treatment type are assigned to each block) are called balanced complete-block or randomized complete-block designs. It is possible to have different numbers of individuals within a block assigned to different treatment groups, or indeed to have treatment groups that do not get experimental units from every block. Such incomplete-block designs are much harder to analyse statistically. If you are forced into such a situation, perhaps by a shortage of experimental units, then our advice is to consult a statistician beforehand.

Q 4.4 Imagine that two scientists need to evaluate the quality of tomatoes grown in an experiment with three different greenhouses and four different treatment types. They are worried that the score a tomato gets might be affected by which of them measures it—can blocking come to their aid?

4.3.2 The pros and cons of blocking

The advantage of blocking is in reducing the effect of intrinsic between-individual variation so that treatment effects are easier to detect. There are drawbacks too. Blocking using a characteristic that does not have a strong effect on the response variables that you eventually measure in order to compare treatments is worse than useless. Hence, *do not block on a characteristic unless you have a good biological reason to think that the characteristic that you are blocking on will have an influence on the variables that you are measuring*. The reason that blocking with an ineffective variable actually reduces your chances of finding a treatment effect is because the statistical test loses a bit of power when you block, because that power is used in exploring between-block differences. Normally this price is worth paying if blocking leads to reduced between-individual variation within blocks; but if it doesn't, you lose out.

Blocking also cannot be recommended if you expect large drop-out rates, such that some treatment groups end up with no representatives from several blocks. This effectively produces an unplanned incomplete-block design, which, as we discussed above, can be very hard to analyse statistically. Generally, the benefits of blocking increase as sample sizes increase, hence we would not recommend blocking in situations where you are constrained to using rather fewer experimental units than you would ideally like. In such cases complete randomization is usually better.

> **?** **Q 4.5** You have a sample of student volunteers using which you intend to explore the effects of four different exercise regimes on fitness. Are there any variables that you would block on?

Don't block on a factor unless you have a clear expectation that that factor substantially increases between-individual variation.

4.3.3 Paired designs

In **paired designs**, we divide the population into pairs and randomly assign the individuals of each pair one to each of two treatment groups.

A common form of blocking is a **paired design**. For example, imagine that we want to look at the effect of an antibiotic injection soon after birth on the subsequent health of domestic cattle calves. We might use sets of twins as our experimental pairs, randomly assigning one calf from each set of twins to get the injection. The attraction of this is that we can perform statistical tests (e.g. paired t-tests or z-tests) that compare within pairs. This has the same benefits as all forms of blocking: eliminating many potential confounding factors, because the two twin calves in a pair will be genetically similar, will be fed by the same mother, and can easily be kept together in order to experience the same environmental conditions. But notice one drawback: since we only perform our investigation on calves that were from a set of twins, can we be sure that our results will also apply to calves that were born alone? This type of extrapolation concern was dealt with in section 3.3.4. Paired designs need not only use sets of twins. For example, a study looking at sex differences in parental feeding in seabirds might compare the male and female partners within a number of breeding pairs. Comparing within pairs controls for variation due to factors such as number of chicks in the nest and local foraging environment, both of which might be expected to contribute considerably to variation in feeding behaviour.

Paired designs can be attractive for similar reasons to blocking generally, but be careful to think about whether your choice of pairs restricts the generality of conclusions that you draw from your sample.

4.3.4 How to select blocks

In section 4.3, we said that it did not matter much how you divided subjects into blocks. For example, if you had 80 subjects and four treatment groups, you could go for 20 blocks of 4, 10 blocks of 8, or 5 blocks of 16. Let's see if we can give you some clearer advice.

First of all, don't have any more blocks that you can justify if you are blocking on something that must be scored subjectively. For example, say the subjects are salmon parr and you are blocking them according to an expert's opinion on their state of health from external examination: given as a score on a 1–20 scale. In this case, it might be hard to convince yourself that those in the 18th and 19th blocks out of 20 are really different from each other. That is, it may be that the subjectivity involved in scoring fish on this fine scale adds noise to your study. If this is a valid concern, then you should use fewer blocks.

If you are ranking on something more objective, like the length of the parr, then you may as well go for as many blocks as you can, in order to reduce variation within a block due to length. However, our interpretation of the analysis of blocked experiments can be complicated if there is an interaction between the experimental factor and the blocking variable. For example, if we are interested in how quickly salmon grow on four different diets, straightforward interpretation assumes that although diet and initial length both affect growth rate, they do so separately. To put this another way, it assumes that the best diet for large fish is also the best diet for small ones. This may not be true. In order to explore whether it is or not, you need several fish on a given diet within each block. Hence, if you are concerned about interactions, and in general you often will be (see Box 4.1), then adopting 10 blocks of 8 would be a better choice than 20 blocks of 4.

In general, make sure that you have replication of a treatment within a block.

4.3.5 Covariates

Let's return to our experiment where we compare the running speeds of greyhounds given different diets. Recall that we expect that differences in age will be an important source of variation between individuals' running performance. An alternative to blocking on the characteristic 'age' would be to record the age of each dog alongside its diet and speed. In our statistical analysis, we could then attempt to control for the effect of age, and so make any difference between the diets more obvious. Statistically, we do this by performing an ANCOVA (analysis of covariance) rather than an ANOVA. There are two requirements for such an analysis of covariance that are important to bear in mind.

The first requirement is that the covariate is a continuous variable such as age. Continuous variables are measured (such as height, weight, temperature) and can take a great range of values and have a natural order to them (for example, 3.2 seconds is shorter than 3.3 seconds, and 5 metres is longer than 3 metres). In contrast, discontinuous (or discrete) variables are categories. They may only take limited values, and need not (but can) have any natural order to them. Typical discontinuous variables might be the colour of a seaweed, classified as red, green or brown (no natural order), or the size of a tumour, classified as small, medium or large (natural order). The distinction is not always clear cut; while classifying dogs as 'young', 'middle-aged' or 'old' treats age as a discontinuous variable, the actual age of a dog in years might be regarded as discrete if only measured to the nearest year, but continuous if we allow fractions of years. See Statistics Box 4.1 later in the Chapter for more on types of variables. If we did not know the exact ages of the dogs, but could identify them as one of three types, i.e. 'young', 'middle-aged' and 'old', then (although alternatives to ANCOVA are available that allow use of a non-continuous covariate) a blocked design is the best option, with the dogs being split into these three blocks.

The second important assumption of ANCOVA is that we expect some form of linear relationship between running speed and age, so that speed either increases or decreases linearly with age. If we actually expected speed to increase between young and middle-aged dogs, but then decrease again as dogs become old, or to follow some other non-linear relationship, then a blocked design will be far more straightforward to analyse. Whether age is treated as a blocking factor or a covariate, your statistical test will be most powerful if the age distributions of the dogs on the different diets are similar.

There is no reason to restrict yourself to one covariate. As a general rule, if there are characteristics of samples that you can measure and that you think might have a bearing on the response variable that you are actually interested in, then measure them. It's much better to record lots of variables that turn out to be of little use, than to neglect to measure a variable that your subsequent thinking or reading leads you to believe might have been relevant (but bear in mind our advice on focused questions in Chapter 2). Once your experiment is finished, you can't turn the clock back and collect the information that you neglected to take first time around.

Recording covariates can be a valid alternative to blocking, but blocking can often be more powerful unless your system satisfies two important assumptions about linearity of response and the continuous nature of the covariate.

4.4 Within-subject designs

The experimental designs that we have discussed so far in this chapter have involved comparing between individuals in different groups, with each group being assigned a different treatment. It is no surprise that

In a **within-subject** design (sometimes called a **cross-over** or **repeated-measures** design), experimental subjects experience the different experimental treatments sequentially, and comparisons are made on the same individual at different times, rather than between different individuals at the same time.

these designs are sometimes called between-groups or between-subjects or independent-measures designs. The alternative is a **within-subject design**, in which an individual experimental unit experiences different treatments sequentially and we compare between measures from the same individual under the different treatments. We have already come across one within-subject design in the context of longitudinal studies (section 3.3.6). Here we explore them in an experimental context.

4.4.1 The advantages of a within-subject design

Imagine that you want to investigate whether classical music makes hens lay more eggs. Off-beat we admit, but you must be getting bored with thinking about feeding tomato plants by now! You could do a simple randomization experiment, taking a number of henhouses and randomly allocating houses either to the treatment group with the classical music or to the control group. One aside here is deciding what the control treatment should be. Should this be a treatment where the sound delivery system is installed but never used, or should it be one where the chickens are exposed to Country & Western music, or should it be one where the sound system delivers various non-musical noises (say a tape-recording of conversations)? This depends on the question being asked, and of course, it might be most interesting to run all three control groups. Anyway, simple randomization is going to require you to have a lot of henhouses. This will be particularly true if henhouses tend to vary in positioning or construction so that we expect inter-house variability to be high.

Q 4.6 If you were carrying out this randomized experiment and had 24 henhouses at your disposal, what control group or groups would you use? The henhouses are of similar construction and are filled with similar hens. The houses are arranged in a large field in two rows of 12.

Q 4.7 Would you use blind procedures in the experiment of the last question?

Q 4.8 Would you be interested in blocking in the experiment above? If so, what factor would you block on?

Q 4.9 Why do we need to allocate henhouses into two groups, each of which gets the treatments in the opposite order?

One potential solution to this problem is to use houses as their own controls, so that you are comparing the same house under different treatments. Let's consider the simple case where you are just using the no-noise control. You should randomly allocate houses to one of two treatment sequences. The first receives classical music for 3 weeks then no music for 3 weeks. The other receives no music for 3 weeks then classical music for 3 weeks. We can then statistically compare the numbers of eggs laid in the same house under the two different regimes, using a paired *t*-test. Because we are comparing within (rather than between) houses, then between-house variation is much less of a concern, and we should be able to get clearer results than from a fully randomized design with the same number of houses.

Comparing within individuals removes many of the problems that between-group designs experience with noise.

4.4.2 The disadvantages of a within-subject design

There are three potential drawbacks to within-subject designs: order effects, the need for reversibility and **carry-over effects**.

Carry-over effects occur when a treatment continues to affect the subsequent behaviour of experimental subjects even after that treatment is no longer applied.

Order effects

Notice that in the henhouse experiment we need two groups: one that experiences the music first and then the control treatment, and another that experiences the two treatments in the reverse order. We need two groups to avoid time being a confounding factor. If the laying rate of the hens changes over time for some reason unrelated to music, then this would confound our results if all the houses were exposed to the music treatment first followed by the no-music treatment. By making sure that we have equal numbers of hen houses experiencing all possible orderings of the treatments (there are only two orderings in our case) we are **counterbalancing**; the situation becomes more complex with larger numbers of treatments (see section 4.4.4).

Counterbalancing in a within-subject design involves allocating at least one experimental unit to every possible sequence of the treatments. If there are N treatments then the number of sequences is $N! = N*(N - 1)*(N - 2)* \ldots *1$.

Reversibility

The within-subject design for the henhouse experiment only works because we can measure non-destructively (we don't have to harm a bird to measure how many eggs it has laid). If the question was whether classical music increases the wall thickness of the heart, then chickens would probably have to be killed in order for this to be measured, and such a cross-over design would not be practical.

More generally, we need to be able to reverse the condition of subjects so that individuals can be exposed to one treatment after another in a meaningful way. For example, it would not be practical to compare the effectiveness of different teaching techniques on how quickly young children develop reading skills using a within-subject design. Once the child has learnt to read using one technique, we cannot wipe away their reading ability so as to measure how well they re-learn under the second technique. Once they have learnt the first time, we cannot take that away. So comparison of the teaching techniques would require a between-groups design rather than a within-subject design.

? **Q 4.10** What if, rather than having the two counterbalanced groups suggested above, we had one group that experienced the experimental treatment for one half of the time and the control treatment for the other half, and another group that experienced the control treatment throughout? Wouldn't this allow us to check for temporal effects?

Within-subject designs are not suitable for situations where a subject cannot be returned to the condition it was in before the treatment started.

Carry-over effects

It may be that the effect of the classical music on the hens' laying rates persists for several days after the music is discontinued. Alternatively or additionally, any disturbance associated with changing the regime may adversely affect the birds. For these reasons it is worth looking at your raw data to see if something unusual appears to be happening at the start of each regime, after which the subjects settle down to a more settled pattern of behaviour. If this is occurring, then it may be prudent to discard data from the period immediately after a changeover. For example, in your chicken study, you might decide to apply each treatment as a 3-week block, but only use data from the last 2 weeks of each block. Within-subject designs are not particularly useful when carry-over effects are likely to be strong and/or difficult to quantify. For example, they are uncommon in drug trials. One possible solution is to introduce a period of normal conditions between the two regimes, to minimize carry-over effects; these are often called washout periods. This is difficult to imagine in our henhouse example, where the control situation is close or identical to normal conditions, but if we were using the conversation control for our hens, then including a week of no stimulus between the two regimes might help to remove carry-over effects. Although this costs you an extra week of husbandry, it may avoid you having to delete data points as unrepresentative.

Make sure that you leave sufficient time for a treatment to stop having an effect on subjects before applying the next treatment and taking measurements of the effect of the new treatment.

4.4.3 Isn't repeatedly measuring the same individual pseudoreplication?

Within-subject designs necessarily involve experimental subjects being measured more than once—why does this not suffer from the problems of pseudoreplication discussed in Chapter 3? The important difference is that in a within-subject design a subject is measured more than once, but there is only one measurement made of each individual under each experimental condition (in the henhouse case, once with music and once without). We then compare between the measurements within the same individual. The problem of pseudoreplication only arises if we measure a subject more than once under the same conditions, and then try to use these data to compare between individuals under different conditions. If you are finding this distinction confusing, Box 4.2 explores the idea further.

Repeatedly measuring the same individual within a well-planned within-subject design is not pseudoreplication.

BOX 4.2 **Within-subject designs and pseudoreplication**

If you are still having problems with the idea of pseudoreplication and within-subject designs, you might find that the following helps your intuition. Imagine we want to compare the running speed of greyhounds on different diets. We have 20 dogs and we measure the speed of each twice, once after a month on the standard diet, and once after a month on a high-protein diet (obviously some dogs get the normal diet first, while others get the high-protein diet first). As this means we have twice as many measurements as we have dogs, surely this is pseudoreplication? To understand why it is not, you need to think about what we are actually interested in in this study.

While we have two measures of speed for each dog, what we are actually interested in is the difference between these measures when each dog has been fed differently. If we wished, we could even turn these measures into a difference score by subtracting the speed on the high-protein diet from the speed on the normal diet for each dog in turn. Doing this makes it immediately clear that what looks like two measures of each individual in our study can in fact be thought of as a single measure of change in a dog's performance. And if we only have a single measure per dog of the quantity we are interested in (i.e. difference in speed), we are not pseudoreplicating. Now while it is true that the situation may become harder to visualize for more complex within-subject designs, as long as we analyse our experiment appropriately to take account of the within-subject measures, the same general principles apply.

If the experiment had been done in a different way, with 10 of the dogs being fed the normal diet, and 10 being fed the high-protein diet, and then each dog had been measured twice, then the two measures on each dog would have been clear pseudoreplicates, and we should combine measurements to get a mean score for each dog before doing any statistical analysis.

4.4.4 With multiple treatments, within-subject experiments can take a long time

OK, lets consider the chickens and music study a little further. Imagine that you are going to use all three controls as well as the treatment group. The straightforward design would be to randomly allocate individuals to different groups, each of which experiences the four regimes in a different order. Now you can see a drawback, especially if we include washout periods: this experiment is going to take a very long time. There may be practical reasons why this is inconvenient. Further, the longer the experiment goes on, inevitably, the greater the problem of drop-outs will be.

There may also be ethical considerations. This is unlikely to be a problem for your musical chickens, but if the experiment involves stressful manipulations of animals, then single manipulations on a larger number of animals may be a better option than repeated manipulations on individuals. This is a difficult decision to make, but it's a biological one, not a statistical one, and you cannot avoid it.

You can shorten the length of the experiment by using an incomplete design, where each individual only experiences some (but not all) of the treatments. However, by now you will be getting the picture that we advise against incomplete designs if at all possible, because they are hard to analyse and certainly beyond the scope of this book. But sometimes they will be unavoidable for practical reasons, and in this case we recommend that you consult one of the texts in the Bibliography and/or seek advice from a statistician.

Within-subject designs with several treatments can take a long time to complete.

4.4.5 Which sequences should you use?

With the four treatment groups of the henhouse experiment, there are 24 different orders in which a single henhouse experiences all four treatments. If you don't believe us, write them out for yourself. Until now, we have recommended counterbalancing, where we make sure that all possible combinations are applied to at least one experimental unit. This was not a problem when we had only two treatments, but with four treatments we would need at least 24 henhouses. If we had five treatments, that number would go up to 120, and up to 720 for six treatments. We can see that counterbalancing quickly becomes unfeasible if the number of treatments is anything other than small. In such situations the normal approach is to randomize the order in which individuals experience the different treatments. That is, we select a random permutation of the treatments for each experimental unit. We must select a random permutation independently for each experimental unit. Each unit then experiences treatments in the order defined by their permutation. This will generally reduce the adverse impact of order effects, but is not as robust as counterbalancing.

Both counterbalancing and randomization are methods of controlling for any order effects that may be present in your experiment. The advantage of counterbalancing is that your statistical analysis can test to see if order effects matter in your experiment as well as controlling for them. Randomization does not allow you to test for order effects in your statistical analysis, so you will not find out how strong they are, although the randomization methodology should help to avoid them confounding your experiment.

If the number of treatments is low (four or less), then counterbalancing should generally be adopted; for a larger number of factors, randomization is a valid alternative.

4.5 Split-plot designs (sometimes called split-unit designs)

The term 'split plot' comes from agricultural research, so let's pick an appropriate example. You want to study the effects of two factors (the way the field was ploughed before planting and the way pesticide was applied after planting) on cabbage growth. Three levels of each factor are to be studied. We have at our disposal six square fields of similar size.

The full randomization method for conducting this experiment would be to divide up each field into, say, six equal sections then randomly allocate four of the 36 sections to each of the nine combinations of ploughing and pesticide application (see Figure 4.4 for a picture of what the experiments would look like). The split-plot method would be to allocate two fields at random to each ploughing type, and then plough entire fields the same way. We would then take each field and randomly allocate two of the six parts to each pesticide treatment. Surely this is a worse design than full randomization—why would anyone do this?

The answer is convenience, and sometimes this convenience can be bought at little cost. It is awkward to organize for different parts of a field to be ploughed in different ways; our split-plot approach avoids this nuisance. In contrast, applying pesticides, which is done by a person rather than a machine, can easily be implemented in small areas. The price we pay for adopting a **split-plot design** is that the split-plot design is much better able to detect differences due to pesticide use (the **sub-plot factor**) and the interaction between herbicide and ploughing, than it is to detect differences due to ploughing (the **main-plot factor**). This may be acceptable if we expect that the effects of ploughing will be much stronger (and so easier to detect) than those of pesticide application, or if we know that ploughing will have an effect but what we are really interested in is the interaction between ploughing and pesticide use.

In a **split-plot design**, we have two factors and experimental subjects that are organized into a number of groups. For one factor (the **main-plot factor**) we randomly allocate entire groups to different treatment levels of that factor. We then randomly allocate individual experimental subjects within each of these groups to different levels of the other factor (**the sub-plot factor**).

Split-plot designs are not restricted to agricultural field trials. Imagine exploring the effects of temperature and growth medium on yeast growth rate. The experimental units will be Petri dishes inside constant-temperature incubators. It would be natural to use temperature as the main-plot factor in a split-plot design, assigning a temperature to each incubator and then randomly assigning different growth media to the dishes inside individual incubators. Be careful here not to forget about

Full randomisation

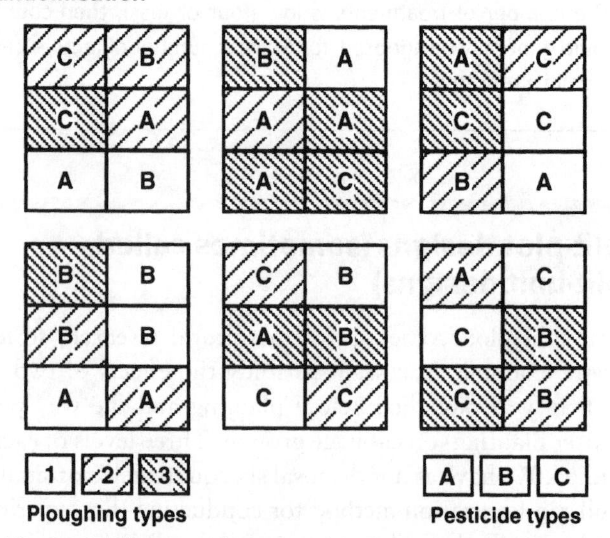

There are 3 × 3 = 9 different combinations of the two treatments. Each of the six fields is divided into six sections, making 36 sections in total.

To allocate plots to treatment combinations we use simple randomization.

For example, we could number all the sections (1–36) then draw the numbers from a hat. We would allocate the first combination to the first four numbers drawn, the next four to another different combination and so on until each of the nine combinations had been allocated to 4 randomly placed sections.

Split plot

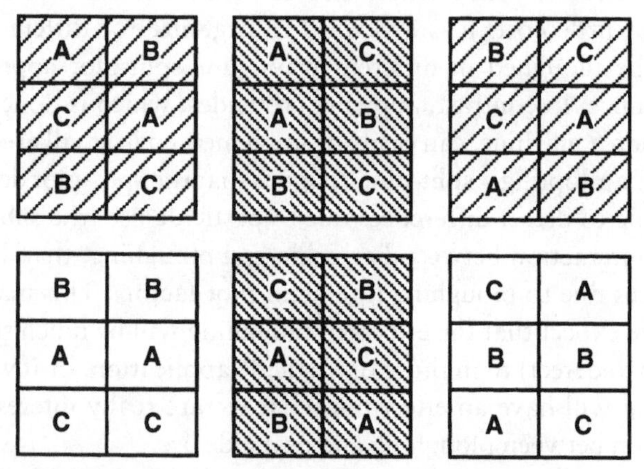

We begin by randomly allocating two complete fields to each of the three ploughing types. That is, all six sections of any given field are each ploughed in the same way.

We then take each field in turn and for each field we randomly allocate the three pesticide treatments among the six sections, such that each field has two sections given each pesticide treatment.

Figure 4.4 Comparison between fully randomized and split-plot designs.

replication. If you care about the effect of temperature at all, then you need replication with more than one incubator for each temperature. It is a very dangerous argument to claim that two incubators are absolutely identical, and so any differences in growth between them can only be due to their different temperature settings. Even if you bought them from the same supplier at the same time, it could be that the difference in their positions within the room means that one experiences more of a draught when open than the other, or one of them may have antibacterial residues from a previous experiment that the other doesn't have, or one could have a faulty electrical supply that leads to greater fluctuations in temperature. Be very, very careful before claiming that two things in biology are identical (as they almost never are), and always replicate.

Q 4.11 We've argued above that replication is important. What would you do in the experiment described above if you only have access to three incubators and you want to explore three different levels of temperature?

Split-plot designs should only really interest you in an experiment with multiple factors where one factor can be allocated easily to groups of experimental units, but is practically awkward to assign to individual experimental units.

4.6 Thinking about the statistics

In any study, the experimental design that you choose and the statistics that you use to analyse your data will be tightly linked. Now that we have outlined some of the most common experimental designs that you will come across, we want to finish this chapter by demonstrating this link with an example.

Suppose that you are in charge of a study to evaluate whether a range of herbal extracts are of any use as antibacterial agents. To do this you decide to measure the growth of *Escherichia coli* bacteria on agar plates containing the different extracts, and also on control plates. You have two research teams at your disposal working in separate labs, and nine extracts to test. How do you proceed?

Sensibly obeying our advice in section 1.3, you decide to think about the statistical tests that you will use to analyse your data before you start collecting any. Your study has two factors. First there is the factor that you are interested in, the different extracts. However, there is also a second factor in this study, which is the lab a particular assay is carried out in. While you are not specifically interested in any small differences in the conditions in the two labs, you would like to include them in the analysis for two reasons: (1) to convince yourself that the results are really not affected by differences between labs, and (2) to control for any small differences between labs, thus making any differences between the extracts easier to detect. Thus your study will be based around a two-factor design.

After discussion with a statistician (or reading a statistical textbook), you decide that the type of analysis appropriate for this study is a two factor analysis of variance (or two-way analysis of variance). Do not worry if you have never heard of this test before, and have no idea how to actually do it. In fact, in these days of user-friendly computer programs, the actual doing of the test is the easy part; it is making sure that your data are appropriate for your tests and vice versa that can be tricky. The important thing from our point of view is that this test, as with all statistical tests, has a number of assumptions that must be met if it is to be used appropriately. One important requirement is that, for each combination of factors, you need at least two measurements. In your case this means that for every extract (and a suitable control), you need to make at least two experimental measurements of growth rate in each of the two labs. So if you had planned to get one lab to test five of the extracts and the other to test the remaining four, think again. A second requirement is that the data collected must be a real measurement (i.e. measured on an interval or ratio scale; see Statistics Box 4.1 for details), not an arbitrary score. So if you chose to measure the average colony diameter of 100 randomly chosen colonies per plate, the analysis would be appropriate, but if instead you simply assessed growth on the plate on a five-point scale, based on categories such as 1 = no growth, 2 = very little growth and so on, you could not use this analysis. There are many other requirements of this test, and while we will not go through them here one by one, that is exactly what you would do if you were in charge of this study. If you can meet all the requirements, all well and good, but what if you can't? First, you need to consider the consequences of breaking a particular assumption. In general, failing to meet the requirements of a test will reduce your confidence in the result (and give ammunition to the Devil's advocate). Failing to perfectly meet some assumptions may have little effect on the outcome of the test, and in this case you might decide to proceed with extreme caution. However, this will often not be the case, and failure to meet many requirements will entirely invalidate the test (which is not to say that you won't be able to get your computer to carry it out, just that the results would be untrustworthy). Detailed discussion of these different situations is well beyond the scope of this book, and if you find yourself in this situation we would strongly advise you to read a good statistics book such as one of the ones in the Bibliography, or to speak to a statistician. If, after this, you decide that you cannot meet critical assumptions, you are left with two options: to either redesign the study to meet all the assumptions or look for another suitable test with different assumptions that you can meet (bearing in mind that such a test may not exist, or may be considerably less powerful than your original choice).

STATISTICS BOX 4.1 Types of measurement

Different statistical tests require different types of data. It is common to split data into the following types.

Nominal scales: a nominal scale is a collection of categories into which experimental units can be assigned. Categories should be mutually exclusive, but there is no order to the categories. Examples include species, or sex.

Ordinal scales are like nominal scales, expect now there is a rank order to the categories. For example, the quality of second-hand CDs might be categorized according to the ordinal scale: poor, fair, good, very good or mint.

Interval scales are like ordinal scales, except that we can now meaningfully specify how far apart two units are on such a scale, and thus addition or subtraction of two units on such a scale is meaningful. Date is an example of an interval scale.

Ratio scales are like interval scales, except that an absolute zero on the scale is specified so that multiplication or division of two units on the scale becomes meaningful. Ratio scales can be either continuous (e.g. mass, length) or discrete (number of eggs laid, number of secondary infections produced).

Our advice is that you should try and take measurements as far down this list as you can, as this gives you more flexibility in your statistical analysis. Hence, if you are recording the mass of newborn infants, then try to actually take a measurement in grams (a ratio measurement) rather than simply categorizing each baby as 'light', 'normal' or 'heavy', as this will increase the number and effectiveness of the statistical tools that you can bring to your data. Of course, by taking our advice and deciding what statistical tests you will carry out on your data before you collect it, you will already know the type of data you require.

Once you have decided on the test that you will use, you can also begin to think about the important questions of the power of the experiment you are proposing, or the kinds of sample sizes that will be required to make your study worth doing. Unless you know how you will analyse your data in advance, it is impossible to make sensible estimates of power (since the power will depend on the test that you choose; see section 3.5.2 for more on power).

While these procedures might seem very long-winded and slightly daunting, especially if you are not yet confident with statistics, we hope that you can now begin to see clearly how thinking about the statistics when you think about your design, and long before you collect data, will ensure that you collect the data that allow you to most effectively answer your biological question.

Think about the statistics you will use before you collect your data.

Summary

- Many experiments require a control group to act as a reference against which experimentally manipulated individuals can be compared.
- Care must be devoted to designing the most effective sort of control.
- Think very carefully before deciding that a historical reference can be used instead of a concurrent control.
- Aim for balanced designs, with the same number of experimental units in each group.
- Multi-dimensional designs allow us to explore interactions between factors, but interpreting such interactions must be done with care.
- In fully randomized experiments, all experimental subjects are randomly assigned to treatment groups without any restriction.
- Simple, fully randomized design is not very powerful in cases where random variation between individuals is high.
- If a given variable is expected to introduce significant unwanted variation to our measurements, then we can control that variation by blocking on that variable.
- In paired designs, we divide the population into pairs and randomly assign individuals within each pair to two different treatment groups, such that the two individuals in the pair end up in different groups. This can be seen as an extreme form of blocking.
- In a within-subject design, experimental subjects experience the different experimental treatments sequentially, and comparisons are made on the same individual at different times, rather than between different individuals at the same time.
- A within-subject design is an effective means of avoiding high between-individual variation masking interesting effects.
- In a split-plot design, we have two factors and experimental subjects that are organized into a number of groups. For one factor (the main-plot factor) we randomly allocate entire groups to different treatment levels of that factor. We then randomly allocate individual experimental subjects to different levels of the other factor (the sub-plot factor).
- There is no theoretical attraction to a split-plot design over complete randomization; however, it may sometimes be simpler to implement, and this practical convenience can (under some circumstances) be bought cheaply.
- Thinking about the statistics that you will carry out on your data before you collect those data will ensure that your design is as good as it can be.

Taking measurements

5

If there is an over-riding message in this book, it is our championing of the need for carrying out preliminary studies before embarking on your main experiment. Chapter 2 covered several of the benefits of pilot studies. Another important benefit is that a pilot study gives you a chance to improve the quality of your recording of results. Your data are extremely valuable. You will have spent a lot of time designing and carrying out your study, so you owe it to yourself to collect the best data you possibly can. However, there are numerous ways in which inaccuracy or bias can creep into your studies if you are not careful, even with the simplest of measurements. Here we will illustrate this with a number of examples of increasing complexity and suggest ways in which common pitfalls can be avoided.

- We begin by illustrating the importance of checking the performance of measuring instruments through a calibration process (section 5.1).
- We then discuss the dangers of imprecision and inaccuracy (section 5.2).
- People are not machines, and their performance can change over time (section 5.3).
- Similarly, two people can record the same event differently (section 5.4).
- There are pitfalls to avoid when dividing continuous variables into categories (section 5.5).
- When observing anything, we must also consider whether we are also affecting it (section 5.6).
- We present some tips for effective data recording (section 5.7) and for using computer technology (section 5.8).
- We discuss measuring experimental units appropriately so as to avoid floor and ceiling effects (section 5.9).
- We then discuss ways of avoiding the accusation of bias in your recording of data (section 5.10).
- We finish with a few final thoughts on the particular challenges of taking good measurements of humans and animals in a laboratory setting (section 5.11).

5.1 Calibration

No matter how simple the measurement appears, there is always reason to be careful in order to avoid obtaining biased or inaccurate results. Take the simplest possible case, where you are simply reading the mass of objects that are placed on a set of scales with a digital readout. Surely nothing can go wrong here? Well, do you know that the scales are accurate? Would it be possible to check by measuring some items of known mass? It is unlikely that your scales are faulty, but it takes little time to check. It would do no harm to do so at the end of your series of experimental measurements as well as at the beginning, just in case they started off fine but something went wrong during your study. This is not overkill—imagine that your experiment involves you carrying out a measurement once a day for 14 days. If you leave the scales on a laboratory bench, then it is not outwith the bounds of possibility that someone could knock them off the bench then stealthily return them to the bench without telling you, or they could drop something heavy on them, or spill a can of soft drink on them. Hence, there is no harm in checking a measuring instrument by getting it to measure something to which you already know the answer, a process called **calibration**. By taking a little extra time to incorporate this into your experiment you will have much improved confidence in what you have measured.

Calibration involves checking or setting a measuring instrument by using it to measure items whose measurements are already known with certainty.

You could reduce the problem of a fault developing during your set of measurements by saving your samples and measuring them immediately one after the other at the end of the 14 days. However, be careful here, this means that some samples will be stored for longer than others. If weight is likely to change during storage, then this could introduce a confounding factor into your experiment. You should see devising a protocol for measuring as part of your experimental design and produce a design that best suits the biology of your system.

Of course, it goes without saying by now that randomization is important in measuring. Imagine in the example above that we have seven samples from a control group and seven samples from a treatment group. These fourteen samples should be measured in a random order (or possibly in an order that alternates control and treatment samples). This avoids time of measurement becoming a potential confounding factor. Why should this matter? First of all, it might be that the set of scales change over time (perhaps as the room heats up during the day) or that the care with which you handle the samples and place them on the scales deteriorates (as it approaches lunchtime). By randomizing, you are not only removing this potential confounder, you are removing any perception that it might be important for your results. If these measures seem excessive, then think of it in terms of satisfying the Devil's advocate mentioned throughout this book.

Calibrate your instruments of measurement and avoid introducing time of measurement as a confounding factor.

5.2 Inaccuracy and imprecision

Imagine that you are measuring the times that it takes students to run 60 metres using a stopwatch with a traditional clockface rather than a digital readout. The students have been subjected to two different types of fitness regime, and your experiment aims to explore whether these regimes differentially affect sprint speed.

The clockface introduces a greater need for interpretation by you than a digital readout does. You have some discretion in how you read off the time, since the second hand will not stop exactly on one of the sixty second-marks around the face. You must then either round up or down to the nearest second or estimate what fraction of a second beyond the last full section the watch is indicating. Our first point is that a digital readout takes away this source of potential **imprecision**. However, if you must use an analogue watch, then make sure that you establish clear rules for how you estimate the time that you record from the watch.

This seems a good point to introduce the ideas of inaccuracy and imprecision. Your judgement in estimating a time from the stopwatch readout introduces a potential source of error. This means that there might potentially be some added noise in your data: if four students all took exactly 7.5 seconds to run the 60 metres, you might record these not as {7.5, 7.5, 7.5, 7.5} but as {7.5, 8.0, 7.0, 7.5} just because of your variation in the way you read the clockface. This imprecision adds between-individual variation to the results, but is random in the sense that it doesn't systematically lead to high or low readings. Imprecision is something which (as discussed earlier) we should seek to minimize in our experimental design. Now what if your stopwatch runs fast, so that you actually record times of {9.5, 9.5, 9.5, 9.5}? This is **bias** or **inaccuracy**. If anything this is generally a worse problem than imprecision. If all we are doing is trying to test whether the two groups of students differ in sprint speed then there is not too much of a problem, since the bias caused by the faulty watch will be the same for both groups. However, if we wanted to quantify any difference in sprint speed then we would have problems. Similarly, we have problems if we wish to compare our results

Q 5.1 Imagine an experiment in which you aim to compare measures of the skeletal size of 1-day-old chicks produced by captive zebra finches on two different diets. One of the diets appears to reduce the chick incubation period, hence causing the chicks born to mothers on that diet to be measured on average earlier in the experiment than those on the other diet. There seems no way to avoid time of measurement as a confounding factor. What should you do?

Imprecision adds errors to your measurements, but in a way that is uncorrelated from one measurement to the next. That is, if one measurement is underestimated then the next measurement might be either under- or overestimated. **Bias** or **inaccuracy** also add errors, but in a correlated way. That is, if one measurement is underestimated by a certain amount, then the next measurement is likely to be similarly affected.

to those recorded with a different stopwatch that does not have the same bias. Worse still, there is no way that your statistical analysis will tell you if you have a bias problem; in fact it will assume that you definitely do not. Hence, you can see that there is a very real reason for calibration, as only this will spot bias. It is sometimes possible to correct your data to remove the effect of a bias. Say, in our current example, that we worked out how fast the stopwatch was running, then we could correct our recorded times appropriately, to get the actual times. However, this can only occur if we can detect and quantify the bias. Calibration checks—like those suggested in section 5.1—are the best way to achieve this. Figure 5.1 presents a classic way to illustrate inaccuracy and imprecision, with the results of several shots, all aimed at the bullseye of a target.

There are other areas of our imaginary experiment where imprecision and bias can creep in. Imagine that you stand opposite the finishing line and start the stopwatch when you shout 'Go!'. There could be a tendency for you to press the stopwatch just fractionally before you shout. There is also an element of judgement in deciding when individuals cross the line. What if some dip on the line like professional athletes and others do not? One way to limit this imprecision is to have a clear rule for when the

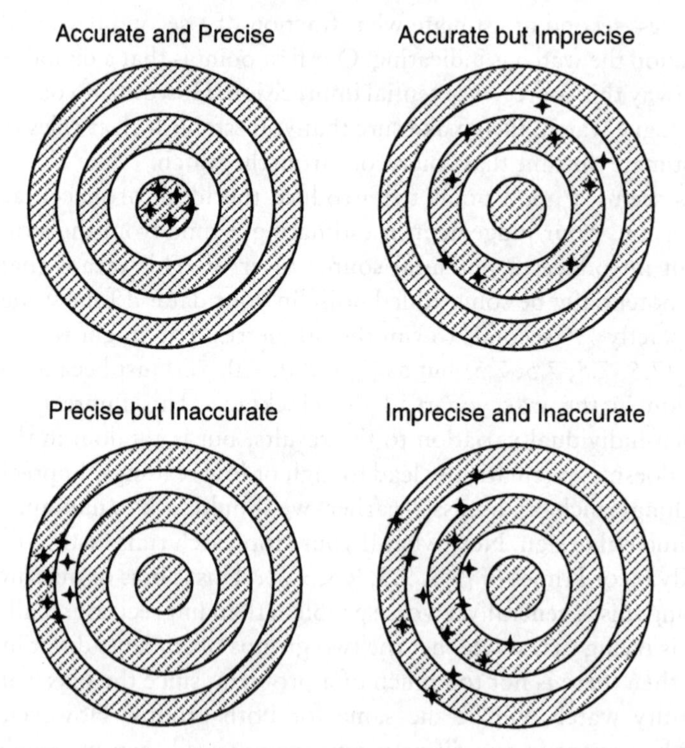

Figure 5.1 The four ways that precision and accuracy can interact, illustrated as the grouping of a number of shots all aimed at the centre of a target.

person crosses the line. Something like, 'I stop the stopwatch when the first part of the person's torso is above the finish line'.

Another potential source of bias in this experiment is that you (standing at the finishing line) are taking it on trust that the people are starting at the start line. In order to reduce their time, individuals might encroach over the line a bit. Further, if individuals see previous students get away with this cheating, then each is likely to encroach a little more than the last. This could lead to completely false results if you have not randomized the order of measuring individuals. Even if you have randomized, it will bias estimation of the size of any effect, since individuals will appear to run faster than they really do. The moral here is to try and design experiments so that there is as little scope for inaccuracy and imprecision as possible. In the running experiment, you might do well to have a colleague checking that the students all start from just behind the line. You should not be worried that students will be insulted by the implied suggestion that they are potential cheats. As ever, you are satisfying the Devil's advocate and removing the *perception* that this might be a potential source of problems in your measurements.

Allied to the last point, this is exactly the sort of experiment where it would be useful if you were blind to the treatment groups that individuals belonged to, until such time as you have measured their run.

All of the above must be making you think that a lot of these problems could be removed if you measured the time electronically, with the clock starting when one light beam (over the start line) is broken and ending when another (over the finishing line) is broken. This would certainly remove some potential sources of error, but remember that it would still be worthwhile calibrating this automated system.

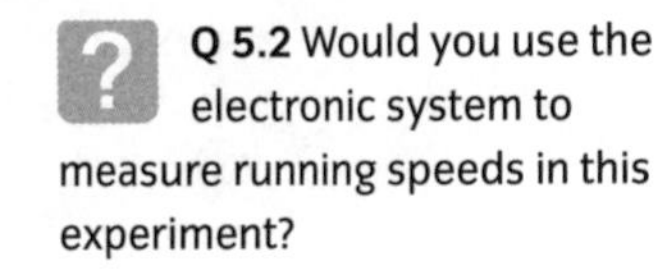
Q 5.2 Would you use the electronic system to measure running speeds in this experiment?

Take time to demonstrate to yourself and to the Devil's advocate that the biases and inaccuracies that exist in your experiment are small enough to have a negligible effect on your conclusions.

5.3 Intra-observer variability

5.3.1 Describing the problem

In our next thought experiment, you are given 1000 passport photographs of 25-year-old males. You are asked to categorize the extent of facial hair in each. You decide to categorize them according to the following scale:

- 0 no facial hair
- 1 a small moustache

- 2 a full moustache
- 3 no moustache and small beard
- 4 full moustache and small beard
- 5 full moustache and substantial beard

You then work through the 1000 passport photos, writing the appropriate number (using the 0–5 scale) on the back of each photo. This takes you 3 hours. We would really be concerned about the quality of these data. Our main concern is **observer drift**. To put it another way, we are concerned that the score a photo gets is influenced by its position in the sequence. This might happen for any number of reasons. First of all, 3 hours is a very long time to be doing this sort of thing without a break. There is a real danger that you got tired towards the end and sometimes wrote '3' even though you were thinking '4'. Also you can subconsciously change the way you define categories over time. Imagine that you get the feeling after doing 50 or so that you haven't given any scores of 2; there is a natural tendency to become more likely to decide that the next few borderline cases between categories 2 and 3 are scored as 2.

Observer drift is systematic change in a measuring instrument or human observer over time, such that the measurement taken from an individual experimental unit depends on when it was measured in a sequence of measurements as well as on its own intrinsic properties.

5.3.2 Tackling the problem

There are lots of things you can do to avoid problems due to observer drift. First of all, you can see that randomization of the order in which you measure is a good idea: if these 1000 individuals belong to two groups, you don't want to do all of one group then all of the other. Now imagine a situation where you know that you'll be given another 1000 photos at some future date. How can you help ensure that your judgement will be the same for that set as this current one? One thing you could do is avoid having a long gap between scoring the two sets if you can. If you cannot, then you could do one (or preferably both) of the following.

1. Try to introduce objective descriptions of the categories or the dividing line between them. For example, a division rule might be 'If the moustache exceeds 80% of the length of the upper lip then it is categorized as "full (category 2)", otherwise as "small (category 1)" '. By reducing the element of subjective judgement involved, you should reduce **intra-observer variability.**

2. Keep a portfolio of examples of different categories. Particularly useful would be examples of cases that are close to the borderline of categories. Looking over these type specimens before starting each further set of measurements should help to avoid drift.

Intra-observer variability is simply imprecision or inaccuracy introduced by human error.

5.3.3 Repeatability

Observer drift is a real problem in lots of situations. However, it has one nice feature—it is really easy to test to see whether you are suffering from it or not. Imagine that you get the second set of photographs 2 months later. Get yourself ready to start on the second set, perhaps by reading your notes from the first set, and looking at the portfolio of type specimens and your objective definitions. Then, instead of starting on the second set, score a random selection of 50 from the first set (taking care not to look at the scores that each got the first time). Now compare your two scores for each of these 50. If they are identical, then it seems that you have little to worry about, and you can move on to the second set confident that whether a photograph is in the first or second set will not influence its score. However, if there are discrepancies, then the nature of these should indicate to you the aspects of your scoring that have changed. For example, you may be tending to see beards as fuller now than before.

Statisticians call this procedure of measuring random samples several times and comparing their values a **repeatability study**, for the obvious reason that you are seeing how well you can repeat the measurements you made the first time.

A measurement is highly repeatable if you get the same score each time you measure something, and has low **repeatability** if your scores tend to be inconsistent. Notice that high repeatability suggests low imprecision but tells you nothing about bias. You can be consistent, but consistently wrong!

There are a number of formal ways that can be used to put a figure on how repeatable a measurement is, but we will not go into those here (see Krebs 1999 in the Bibliography if you are interested in finding out more). However, even simply comparing the data by eye (maybe by plotting a graph of the first measurement versus the second) will alert you to any major problems. Of course, you can look at the repeatability of any measurement, not just category measurements like our beard measures. Even in the study discussed above with the digital balance, it might be worth measuring several samples twice, to ensure that the weight measurements are repeatable. Knowing that your measurements are repeatable will give you greater confidence in their quality.

So it is easy to test whether you have a problem with observer drift. However, what can you do if it does appear to be a problem? In the case above, we would recommend going back to the original set, and picking out cases that are intermediate between the two beard types. Have a look at the scores you gave them before. See if you get a feeling for how you were scoring. If you get this feeling, then go and do some other stuff for a while and completely forget about facial hair. After this break, come back and take the 50-photo test again. Don't just look at photos that were borderline between small and full beards, take another random sample and see if your agreement between old and new scores is now perfect or at least much improved. If so, you are probably ready to score the second set; if not, just keep repeating this procedure until you get yourself back into the same (scoring) frame of mind as you were in for the first set. The bad news is that sometimes this won't happen. Sometimes you will just not be able to understand how on earth you gave the first set the scores you did. When this happens, you have little choice but to re-score the first set at the same time as you are doing the second. This means that your original 3 hours were for nothing: you can see that it pays to try and have as objective a scoring system as possible, to reduce the chance of this happening.

Q 5.3 A study looks at domestic cats on different diets. You are a veterinarian blind to the experimental groups that cats belong to, but must measure body condition by taking a body mass and measuring the skeletal size of the cat along their back from the top of the head to the base of the tail. Owners bring their cats in for measurement only once, and you perform this task on four different cats each week for 20 weeks. How can you defend yourself against possible accusations about drift?

Observer drift can easily occur, but it is easy to check if it has occurred, and if you detect it then remedial actions are available. But prevention is better than cure!

5.3.4 Remember, you can be consistent but still consistently wrong

The methods described in the last section of re-measuring and comparing allow you to spot whether you are consistent in your measurements; they do not tell you whether you are consistently right or consistently wrong. Consider a situation where you have to look at 100 X-rays of horses' legs and identify whether or not each shows a fractured bone. If you score all of these, then re-score them all and find that your assessments agree, then this is encouraging; you are not drifting in your assessment. However, is your assessment right? Where there is a degree of subjectivity (and there is a considerable degree generally in looking at X-rays), then there is the danger that you have a consistent bias. Perhaps there is a certain type of shadow that you tend to consistently misidentify as a fracture, or a certain type of hairline fracture that you consistently miss. How can you spot this sort of bias? One approach would be to get someone to make you up a set of X-rays where the answer is known with certainty. You could then score these blind to the correct solutions, before comparing your performance with the correct answers. This will not only test if you are scoring effectively, but will also help you identify the common thread between any cases that you get wrong. Another alternative, if a set of absolutely correct answers is not available, is to get an acknowledged expert to score at least a selection of your X-rays. Comparing your scores with theirs may help to identify situations where you may be scoring incorrectly. If an expert is not available, then use anyone. Comparing your scores with anyone else's will be instructive in identifying the aspects of an X-ray that are likely to lead to problems in scoring. This leads us onto another potential problem: **inter-observer variability**.

Inter-observer variability is error that arises because several observers are used to gather a body of data. Two observers (or more generally two different measuring instruments) are not exactly alike; differences between them can add imprecision and, if you are not careful in your design, bias.

5.4 Inter-observer variability

5.4.1 Describing the problem

If you can score differently on two different occasions, then it should be no surprise that two different people can score differently. This has implications for how you design your data collection. If possible, the same person should take all similar measurements. You should especially avoid situations where one person measures all of one treatment group and another person measures all of another treatment group.

5.4.2 Tackling the problem

Sometimes practical reasons force you into a situation where you must use more than one observer. The key thing here is to try and make as sure as possible that the two observers agree on how they score things. The way to explore this is exactly as we outlined in section 5.3.3, but instead of you scoring the same set of individuals at two different times and comparing, now the different observers score the same set of subjects and compare their results.

Q 5.4 What are the problems with different people measuring different treatment groups?

Compared to intra-observer variability, it is generally easier to combat inter-observer variability, because when two of you disagree on scoring of the same thing, then you can have a discussion on why this has occurred and what you can do to prevent it happening again. As with intra-individual variability, reducing the need for subjective assessment is a sure way to reduce inter-individual variation. Similarly, don't just check to see if you agree with other recorders at the start of the experiment, check at the end too, to see if you have drifted apart. Finally, if you design your experiment correctly, and note down which observer made which measurements, you can even use the techniques that we described under blocking in Chapter 4 to deal with any small inconsistencies between observers (the same applies if you are forced to use more than one piece of measuring equipment in the same study, e.g. several stopwatches). In essence, we can treat each observer as an experimental block (like the greenhouses in section 4.3.1), and as long as each observer has taken measurements in all treatment groups (and we recommend very strongly that you make this the case to avoid problems of confounding) it is reasonably straightforward to control for differences between the observers in the same way as we did with differences between greenhouses. However, as with all things statistical, prevention is better (or at least easier to analyse!) than cure, so avoiding inter-observer variability is worth aiming for.

Try to minimize the number of observers used, but if you must use several observers then adopt our recommended strategies to minimize or allow you to control for between-observer differences.

5.5 Defining categories

Assigning individuals to categories is often at the root of inter- and intra-observer variability. Take something as simple as watching rhino in a zoo, and categorizing each individual's behaviour at 5-minute intervals as 'standing', 'walking' or 'running'. Here are some things to consider.

First, we are glad that you have gone for three categories rather than fifteen. The more categories that you have, the more room for error there is. But you might consider a system where you record each rhino as either 'static' or 'moving' and if they are moving you record the rate of striding. This has the attraction that it gives you more precise information on movement rates and avoids having to make a tricky decision about when a fast walk becomes a run.

If you stick with the three categories, you have got to be sure that you can assign any given situation into one and only one of these categories. So you need to have a clear rule about how to differentiate between standing and walking and between walking and running. Is shifting of the feet while turning on the same spot walking or standing? Ideally all the observers should get together during a pilot study to work this out. They should finish with a clear statement of how categories are defined and how data are to be recorded, then they should try independently observing the same behaviour and see how they agree.

Define your categories as explicitly as you can, and test your categorization scheme in a pilot study.

5.6 Observer effects

As with the hypothetical case of Schrodinger's cat, there are times when the simple act of observing a biological system will change the way it behaves. Imagine you are watching the foraging behaviour of fish in a tank, in both the presence and the absence of a model predator. Unfortunately, although the fish are not particularly fooled by your model predator, the fact that you are sitting over the tank watching them does worry them. As a consequence, all the fish behave as if there was a predator present all the time, and you cannot detect any difference due to the model's presence. These problems are not just limited to behavioural studies either. Suppose you want to know whether rodent malaria induces the same types of fever cycles as seen in humans. The simplest way to do this would be to take mice infected with malaria and measure their temperature at regular intervals and compare this to control mice. However, the mere act of handling mice to take temperature readings is enough to induce hormonal changes that increase the body temperature of the mice. The process of taking the measurement has made it an unreliable measurement.

Phenomena like those described in the last paragraph present real problems when carrying out biological studies. How can we deal with them? One way is to allow the animals to acclimatize. Whenever we take animals into the lab, it is essential to give them time to settle, so that their behaviour and physiology are normal when we carry out our studies. One important part of the acclimatization might be getting used to the observer, or the handling procedures being used. Often observer effects will disappear after a suitable time of acclimatization. If this doesn't work, a next course of action might be to try to take measurements remotely. Our fish in the tank could be watched from behind a screen, or using a video camera. Similarly, temperature probes are available that can be placed on the mouse, and record its temperature at pre-defined intervals, removing the need to handle the animals during the experiment. Such equipment might be expensive, but might be the only way to get good data. It may also provide an ethical benefit if it reduces the stress the animal suffers.

The biggest problem with observer effects is that, by their very nature, it is hard to tell whether you have them or not. How do you know how the fish would be behaving if you were not there watching them? The answer is to try to take at least some measurements remotely and compare these to direct measurements. It might be that you don't have access to enough video recorders to video all of your fish, but if you can borrow one for a few days and carry out a pilot study to compare the way fish behave with and without an observer present, then you might be able to show convincingly that you don't need to video the fish in the main study. In the end, the question of observer effects will come down to the biology of your system, but you should do all that you can to minimize them, and then make it clear what you have done, so that others can evaluate your results.

Q 5.5 One case where observer effects are very possible is when collecting data on humans by interview. How can you best address this?

Make sure that the way you collect information does not affect the variables that you are trying to measure.

5.7 Recording data

Recording data well is an art form. Doing it successfully will come in part with experience, and there are a number of tricks and tips you will pick up along the way. However, in an effort to get you started, here are some of the things that we wished someone had told us when we started. Some of these may sound trite to you; all we can say about the advice below is that it is advice that we have ignored in the past to our cost.

5.7.1 Don't try to record too much information at once

This is particularly likely to happen in behavioural studies. You are only human, and the more different types of data that you try to record at once the more mistakes you will make. Your pilot study will give you a chance to work out shorthand codes and data sheet designs for efficient recording of information. If you really need to record a diverse stream of information from a behavioural study, then videotape it, and then watch the video several times, each time recording a different aspect of the data.

5.7.2 Beware of shorthand codes

While watching a group of chimpanzees, you develop a code such that when an activity starts you speak a two-digit code into a continuously running tape-recorder. The first digit is the individual identifier (A, B, C . . . H) and the second is the type of behaviour:

- E eating
- A aggression
- P play
- G grooming
- V vocalizing
- W watching for predators

Our first point is that it will be much easier to differentiate between 'Delta' and 'Bravo' than between 'b' and 'd', when you come to transcribing your tapes. Make sure that when you come to transcribing the tapes that you can remember what 'W' was. Probably the safest thing is to define the code at the start of each tape that you use, that way the encoded data and the key to the code cannot become separated. The same goes for information on the date, location, weather and whatever other information about the data collection you need to record—generally strive to keep related information together.

5.7.3 Keep more than one copy of your data

Data are irreplaceable, but computers break down and notebooks get left on trains. Transcribe your fieldwork journal onto a computer spreadsheet as soon as possible, keep a backup of the spreadsheet and keep the original notebook. If it is impractical to transcribe it quickly, then photocopy pages regularly and post the photocopies to your mum, or at least keep them far away from your original notebook. If you lose data that you don't have a copy of, that's not being unlucky, it's having your carelessness exposed.

5.7.4 Write out your experimental protocol formally and in detail, and keep a detailed field journal or lab book

Imagine that you have just finished an intensive 3-month field season watching chimpanzees in the wild. You deserve a holiday. After taking 3 weeks off to explore Africa, you come back home. It takes you another couple of weeks to work through the backlog of mail that has accumulated while you were in Africa. Hence, by the time you come to collate and analyse your data, it is 1–4 months since you actually collected those data. Expect to remember nothing! Everything to allow you the fullest possible analysis of your data must be recorded in some form (paper, computer disk or audio tape). You cannot trust your own memory. Record everything that matters, and record it in a full, clear and intelligible form. The test you should use is that your notes should be sufficiently full that someone else with a similar training to you (who has spent the last 3 months looking after your cat while you were in Africa) could take your data and analyse them as well as you could. So don't record the location as 'near the top of Peacock Hill', give the GPS co-ordinates, or mark the spot on a map. Your notes should also include detailed descriptions of the categories you have been measuring, and any other measurement rules you came up with (did you measure arm length with the fingers included or not?). Careful note-keeping also has the added advantages that it makes it far easier for someone else to collect data for you should you become sick for a few days, and allows someone to follow up your research easily in the future.

5.7.5 Don't over-work

A corollary of the last point is that there is a tendency to feel that you only have 3 months to watch these chimpanzees, and so every waking minute not spent watching them is a minute wasted. Striving to collect as much data as possible is a mistake. Quantity is comforting, but this feeling deceives you: quantity is no substitute for quality. We will not mount our hobbyhorse about pilot studies again. However, time spent transcribing, collating and checking the quality of your data is also time well spent, and will let you spot problems as soon as possible. Also, the sooner you transcribe your field notes into a more ordered form (probably by typing them into a computer) the less time there is for them to get lost or for you to forget important pieces of information. Most difficult of all, you must take breaks and do things unrelated to your work (sleeping, eating and bathing are all to be recommended). This is not about admitting that you are weak, it's about admitting that you record data better when you are fresh than when you are worn out. Don't be swept along by a macho culture that says you have got to suffer to collect good data. Tired recorders make mistakes and they don't notice when they are doing something stupid. Working too hard is dangerous and pointless: don't do it!

Q 5.6 How long can you look down a microscope at slides and count pollen before your accuracy begins to deteriorate through fatigue?

Above all else: never, never, never have only one copy of a set of data.

5.8 Computers and automated data collection

Computers are a real boon in data collection. Compared to humans, they are generally more precise, less likely to drift and less likely to get tired or bored. However, their one big drawback compared to humans is a lack of self criticism; they will record rubbish (or nothing) for days on end without giving you any indication that something is wrong. Hence it is essential that you check that your computer is recording what you think it should be. Pilot studies provide an ideal opportunity for this. Keep re-checking during the course of the study.

Computers are great, but don't go thinking that they are fool-proof.

5.9 Floor and ceiling effects

Imagine that you want to investigate the effect of alcohol on people's ability to absorb and recall information. Your design has two groups of ten people. Individuals in the alcohol group are allowed to relax and listen to music for a couple of hours while they consume a fixed number of alcoholic drinks. Individuals in the control group experience the same regime except that their drinks are non-alcoholic. Each individual then watches the same 5-minute film before being given a number of questions to answer about the film they have just seen. The difficult part of designing this experiment is in preparing appropriate questions: they must be challenging but not too challenging. If the questions are so hard that none of the individuals in either group can get them correct then you learn nothing about differences between the groups. Similarly, if all twenty people can answer the questions without difficulty, then again you can learn nothing about the cognitive effects of alcohol. The first of these is a **floor effect** in the jargon, and the second is a **ceiling effect**. Ideally you want questions that are challenging but not impossible for both sober and intoxicated people to answer, so that you can measure differences in the extent of their accuracy in answering all the questions. The best way to avoid floor and ceiling effects is a pilot study.

A **floor effect** occurs when the majority of measurements taken from experimental units are at the lowest value of the possible range; a **ceiling effect** occurs when the majority take the highest value.

A floor effect occurs when all (or nearly all) measurements taken from experimental units are at the bottom value of the possible range. A ceiling effect occurs when all (or almost all) measurements on experimental units are at the top value of the possible range. You should aim to avoid these effects, since you need to record a range of scores between individual experimental units to hope to have any ability to detect differences due to experimental treatments.

For example, say we randomly assign undergraduate volunteers to two different fitness training regimes. We wish to measure some aspect of fitness that we can then use to assess if one regime is more effective than the other. If we use the simple measurement of whether 50 metre sprint speed has improved after 1 month's training, then we are likely to find that all individuals on both regimes can sprint faster after 1 month's training, and so we learn nothing about any differences there might be between the regimes. We have fallen foul of a ceiling effect. If, alternatively, we ask how many months on the training regime are required to reduce sprint speeds by 50%, then we run the opposite danger of setting a task than no-one on either regime can achieve. We fall foul of a floor effect. Instead, we should seek to take a measurement across individuals that would provide considerable between-individual variability. Perhaps we could measure the percentage reduction in 50 metre sprint speed after 1 month's training. We can then use a statistical test to ask how much (if any) of the between-individual variation in this measurement can be explained by the differences in training regime.

Avoid tests of subjects being too easy or too difficult for variation between individuals or groups to be picked up.

5.10 Observer bias

You are always looking to convince the Devil's advocate that there is no way that your recording of data can be susceptible to conscious or unconscious bias. You need to be able to defend yourself against the accusation that you find less growth of your pathogen on plates treated with an antibiotic because that is what you expect to see. The suggestion is that, when there is ambiguity as to whether the pathogen has grown on a plate, your decision to record that plate as positive or negative for the pathogen is biased by your expectation. You can tackle this charge in two ways: by removing the ambiguity and by removing the expectation. You should design your data collection so as to have as little requirement for subjective judgement by the measurer as possible. Hence, instead of having someone simply look at the plates and pronounce whether each is positive or negative for the pathogen, you should decide on some objective measure before measurement begins. In this case we might take digital photographs of the plates, decide on the colour range that codes for the pathogen, measure how much of the plate shows this colour using image analysis software and have a rule for the minimum percentage that signals pathogen growth. This seems like quite a lot of work when you feel that ambiguous plates will occur relatively rarely. We agree: in this case it might be better to tackle the issue of expectation. If the person who determines the pathogen status of each plate does not know which plates have the antibiotic, or indeed that there are even two different experimental groups (antibiotic-treated and untreated control), then they cannot be biased by an expectation that they do not have. This is a blind procedure as discussed in section 4.1.2. Our advice would be to design your data collection techniques to minimize requirements for human subjective judgement. Where appreciable levels of human judgement are required then blind procedures should be implemented whenever possible.

Q 5.7 Sometimes you cannot avoid subjectivity and cannot practically use blind procedures. Imagine that you require to decide on the sanity of convicted murderers in a study that looks at gender differences. There is no objective test for sanity, so you are going to have to rely on the (at least potentially subjective) opinions of psychiatrists. These psychiatrists are probably going to have to interview the subjects, making it very difficult (if not impossible) for them to be blind to the subjects' genders. How can you defend the study against accusations of biased measurement?

Always ask yourself whether the Devil's advocate would see any potential for biased measurements in your study: try to restrict their ammunition as much as you can.

5.11 Taking measurements of humans and animals in the laboratory

For both ethical and practical reasons you must avoid unintentionally stressing subjects. A stressed individual may not produce the same measurements of the variables that you are interested in as an unstressed individual in its normal surroundings. Stress is often caused by unfamiliarity with the surroundings. Build familiarization into your measurement procedures, such that you explain as much as is practical to human subjects before you start measuring them (see section 6.7). Give them a chance to familiarize themselves with any equipment if this is possible, and encourage them to ask questions about anything that's on their mind. For measurements on laboratory animals, animals are commonly moved from their normal accommodation to a test chamber. Both the process of moving and the experience of new surroundings are likely to stress the animal. Solutions to this are going to be highly species-specific, but our advice is to study the experimental protocols of related previous works with a view to minimizing animal suffering and minimizing the danger of getting artefactual measurements because of unintentional stress caused to subjects.

You can be sure that the potential for such abnormal behaviour will be something that occurs to the Devil's advocate: do all that you can to prepare a defence against such concerns.

Summary

- Your design must include consideration of how you'll take measurements.
- Calibrate your measuring instruments (including human observers).
- Randomize measurements.
- Adopt clear definitions to reduce subjective decision-making during measurement taking.
- Watch out for observer drift, intra-observer variability and inter-observer variability.
- Watch out for observer effects, where measuring a system influences its behaviour.
- Recording data effectively is a skill that you must acquire.
- Automated data collection has many advantages, but can go very wrong without you immediately noticing.
- Beware of floor and ceiling effects.
- Be ready to defend your study against accusations of observer bias and/or stressed live subjects.

6 Final thoughts

In the previous chapters of this book, we have outlined the basic tools that you will require to design sensible and informative experiments that will provide valuable data to answer the questions that you are interested in. If you use these tools, you should avoid many of the pitfalls that await the unwary. You will also be able to confidently defend your work against the attacks of the Devil's advocate (not to mention those of your supervisor and colleagues). You will also be better able to spot any flaws in the work of others. In this final chapter we deal with a number of more specialist issues that you may come across at some point in your career. In some cases we will not address the issues in any great detail, but wish to simply make you aware of them so that you can consult a book or friendly statistician should you find yourself in a situation where they might be useful.

- Section 6.1 considers how the different levels for a treatment should be selected.
- Section 6.2 deals with subsampling, when there is a trade-off between the number of independent experimental subjects that you can use and the number of measurements (subsamples) you can make on each subject.
- You should generally always seek to use equal numbers of experimental subjects in each treatment, section 6.3 considers one relatively uncommon but important exception to this rule, where ethical considerations drive you to adopt an unbalanced scheme.
- Section 6.4 presents three occasionally useful alternatives to random sampling: sequential, stratified and systematic sampling.
- Section 6.5 introduces Latin square designs.
- Section 6.6 has further thoughts on interactions.
- Section 6.7 discusses some special considerations particular to dealing with human subjects.

6.1 How to select the levels for a treatment

Suppose that you want to know what effect a fertilizer has on the growth rate of tomato plants, and you have 30 plants at your disposal. How should you spread these plants across fertilizer treatments? One extreme would be to choose the highest level of fertilizer you might reasonably add (say 30 g) and the lowest (maybe 0 g), and run an experiment with just these two treatments, with 15 plants in each treatment. At the other extreme, you might decide to keep the same limits but have many more treatment groups spread between these values. You might choose to give one plant 0 g, one plant 1 g, one plant 2 g and so on, increasing the dose in 1 g intervals until you get to 29 g. Or you could choose to have two plants at each concentration, and have treatments spaced in 2 g units. In fact, there are many ways to carry out this experiment: how do you choose which is best?

The answer to this question, as with so many questions of design, will depend on your knowledge of the biology underlying the system. Suppose that you know, or at least are very confident, that the effect of fertilizer on growth rate increases linearly with concentration (as in Figure 6.1a), and you simply want to know how rapidly growth rate increases with increasing fertilizer (i.e. the slope of the line). In that case, the most powerful design will be to have two extreme treatment groups, one at low fertilizer concentration and one at high fertilizer concentration, and to split the plants in order to have half (15 plants) in each group. This will give you the best estimate of the slope, and also the highest statistical power to detect a difference between these groups (remember from Chapter 3, statistical power increases with the size of the biological effect, so by choosing two extreme treatments, you are maximizing your statistical power).

However, what happens if you don't know that the relationship is a straight line? Maybe the effect of fertilizer increases to a plateau (as in Figure 6.1b) or increases and then decreases (as in Figure 6.1c). In either of these cases, taking only two treatment groups will give you misleading information about the relationship between growth rate and fertilizer concentration. In the first case, you would not be able to see the important biological (and probably economic) effect that, at low levels of fertilizer, a small increase has a big effect on growth rate, while at high levels, the same increase has little effect. In the second case, you might conclude that fertilizer has no effect (or even a negative effect!), even though at intermediate concentrations fertilizer has a big positive effect. The solution is clear: if you don't know the shape of the relationship, and cannot be confident that it is linear, then you need to have intermediate treatment groups to allow you to assess the shape of the relationship. To get the best estimate of the shape of a relationship, you will need to have as many

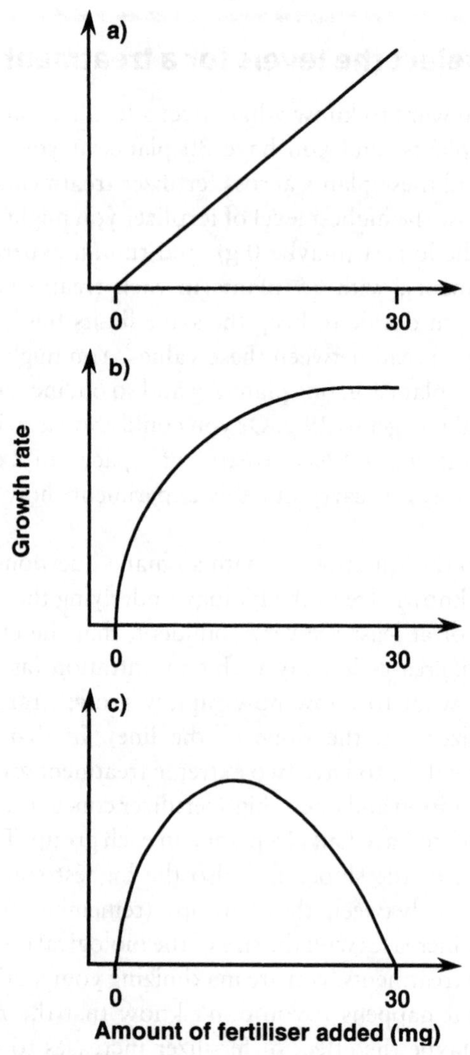

Figure 6.1 Three possible shapes to the relationship between amount of fertilizer added and increase in plant growth rate.

treatment groups as possible, spanning the range that you are interested in, and as evenly spaced as possible.

Even if you are reasonably confident that the relationship is linear, it may be worth hedging your bets to some extent and including one intermediate treatment group. This may slightly reduce your statistical power, but will provide a data set that you can be more confident about.

There are pros and cons to all methods of selecting levels; choose the design that you think will be best for your particular question and system.

6.2 Subsampling: more woods or more trees?

A problem that will crop up in many studies is how to allocate your sampling efforts between two different levels. We will illustrate this with an example. Imagine you are interested in testing the hypothesis:

> The numbers of species of beetle found on trees differ between local coniferous and broad leaf forests.

You consult your map and find 10 possible forests of each type, and you decide that you have the resources and time to examine 100 trees in total. How should you allocate your sampling across forests? One solution would be to choose one forest of each type and then examine 50 trees in each of these forests (selected at random within forests, of course). This would give you the maximum amount of information on each forest, but on the minimum number of forests (only one of each type). If you are interested in general differences between coniferous and broad leaf forests, this would clearly not be the way to proceed. Another alternative would be to sample all 20 forests, which would mean that you could examine only 5 trees at each site. This would give you much less information on each site, but would give you at least some information on the maximum number of forests. Alternatively, you could compromise between these extremes, and sample five forests of each type, examining ten trees in each. Which of the three designs (illustrated in Figure 6.2) is best?

Q 6.1 In the text, we say that if you are interested in uncovering any general differences between coniferous and broad leaf forests, then studying only one forest of each type is 'clearly not the way to proceed'. Can you explain why in your own words?

To identify the best design, we need to go back and think about the purpose of replication. Replication allows us to deal with variation. In this example we have two sources of variation:

- Trees within a single forest will vary from each other in how many beetle species they carry.
- The forests will differ between each other in the average number of beetle species that the trees within them carry. This may occur because of differences between forests in altitude or hydrology, for example.

We need to bias our sampling towards the source of the greatest amount of variation. At one extreme, imagine all trees within a forest have exactly the same number of beetle species. That is, imagine that there is no variation between trees within a forest. In such a situation, we would only need to sample one tree within each forest to get a good estimate of the number of beetle species per tree at that site, and we could concentrate on measuring as many different sites as possible. At the other extreme, what if all the coniferous forests have the same average number of beetle species per tree, and all the broad leaf forests have the same

Three designs for sampling 50 conifer trees

a)

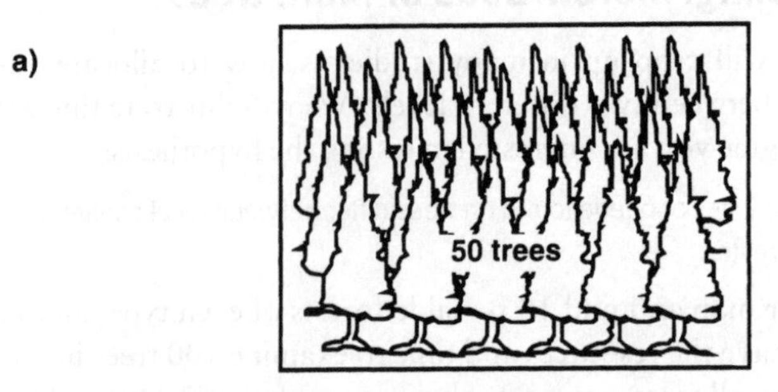

50 trees all from the same forest. **Excellent** information about that forest but no information on other conifer forests

b)

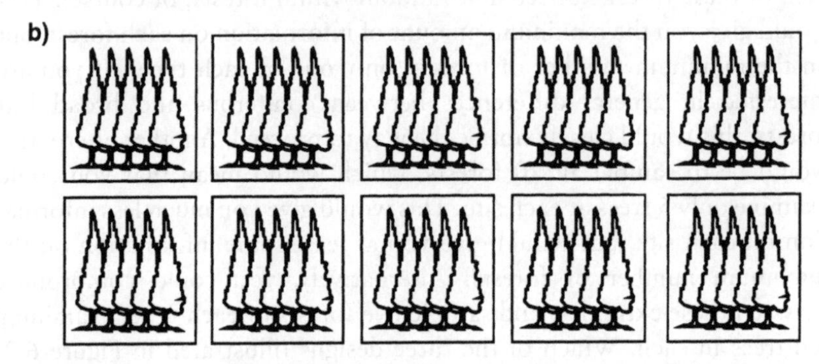

5 trees from each of 10 forests. **Fair** information of a good sample of different conifer forests.

c)

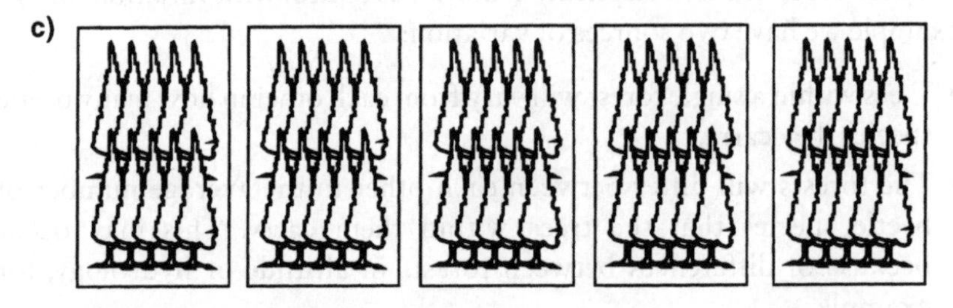

10 trees from each of 5 forests. **Good** information on a fair sample of different conifer forests

Figure 6.2 Three of the possible alternatives for sampling conifer trees.

average number as each other (but different from the conifers), but trees within a site vary (about the appropriate average) in the number of species they carry? In this case there would be little need to sample many sites of each type and we could concentrate on getting better data from each site. Thus, you can see that the answer to the question depends on the biological variation that is present. With the benefit of pilot data, it would be possible to estimate the power of designs with various different sampling regimes to determine which is the best one to use. Such techniques are similar to those discussed for simple designs in section 3.5, but this is not the place to discuss them further. The book by Underwood (1996) in the bibliography has a particularly thorough consideration of these issues. *However, as a good rule of thumb, unless you believe that the variation within independent experimental subjects (forests in this case) will be very large, or you are already sampling many subjects, increasing the number of subjects will generally give you more power than increasing the number of measurements (trees in this case) from a subject.*

Bias your sampling towards getting more measurements at the level where you expect the greatest variation.

6.3 Using unbalanced groups for ethical reasons

As we have emphasized throughout this book, one of the main objectives of experimental design should be to reduce to the absolute minimum any suffering by experimental subjects. Often this will be achieved by using as few subjects as possible, while still ensuring that you have reasonable chances of detecting any biologically significant effect, and in the absence of any other extenuating circumstances this is a good rule of thumb to follow. However, sometimes with a little careful thought, we can do better than this.

Suppose that we wish to carry out an experiment to investigate the effects of some experimental treatment on mice. Our experiment would obviously include two groups to which mice would be allocated at random: one experimental group that would receive the treatment, and a control group that would be kept in exactly the same way, except that they would not receive the treatment. First let's assume that we are limited in the number of mice we can use, to say twenty. How should we divide our mice between groups? To carry out the most powerful experiment, the best way to allocate our mice would be to have ten mice in each group. You will remember from Chapter 4 that an experiment with equal numbers of individuals in each treatment group is referred to as a balanced experiment, or balanced design, and in general balanced designs are more powerful than unbalanced designs. *Thus, if we wish to maximize our probability of seeing an effect in our experiment, we should balance it.*

However, imagine that the treatment that we are applying to our mice is stressful, and will cause some suffering to them. With a balanced design, ten of the mice will experience this unpleasant treatment. It would be better ethically if we could move some of these mice from the experimental group into the control group, and reduce the number of mice that suffer the unpleasant treatment. If we carried out our experiment with five experimental mice and fifteen control mice this would mean that only five mice experience the unpleasant treatment. The down side of this is, of course, that our experiment has become unbalanced, and so will be less powerful than the original balanced design. If the drop in power is not great, we might accept that the change is worthwhile to reduce the number of mice that suffer. However, if the drop in power is too great, it may make the whole experiment a waste of time, as we would have no realistic chance of detecting a difference between the groups. In that case fewer mice will have suffered, but they will all have suffered for nothing. Is there any way that we can both reduce the number of mice that suffer, but maintain statistical power? The answer is that we can, if we can use more mice in the experiment overall.

As you may remember from section 3.5.3, one of the determinants of the power of an experiment is the total sample size of the experiment. We can increase the power of our unbalanced experiment by increasing the total sample size. Practically, this would be achieved by increasing the number of mice in the control group. Thus, we might find that an experiment with five experimental mice and twenty-one control mice has the same power as our original experiment with ten experimental and ten control mice (see Figure 6.3). The experiment has the down side of using more mice in total (twenty-six compared to twenty), but the benefit that fewer of them have to experience the unpleasant experimental treatment. Presumably, even the control mice will experience some stress from being handled and kept in the lab, so is increasing the number of mice that experience the mild stress of simply being in the experiment outweighed by the reduction in the number that experience the very stressful treatment? This will depend on, among other things, how stressful the treatment is and how stressful simply being in the lab is. This is an issue for you as a life scientist: it's not something that anyone else can decide for you (although that doesn't mean that you shouldn't seek advice!).

The details of how to design effective unbalanced designs are beyond the scope of this book, but the paper by Ruxton (1998) in the Bibliography provides further discussion. In principle, it is simply a matter of determining the power of the different possible experimental designs in the same way as was done in section 3.5.2. We mention it here in the hope that by making you aware of the possibility, you will be encouraged to investigate further should the need arise. Of course the same logic can also be used in contexts other than ethical concerns. Imagine that the chemicals used in the treatment are very expensive or the treatment is time consuming: it may pay to carry out a larger unbalanced experiment with fewer individuals in the expensive or time-consuming group.

Control group **Experimental group**

a)

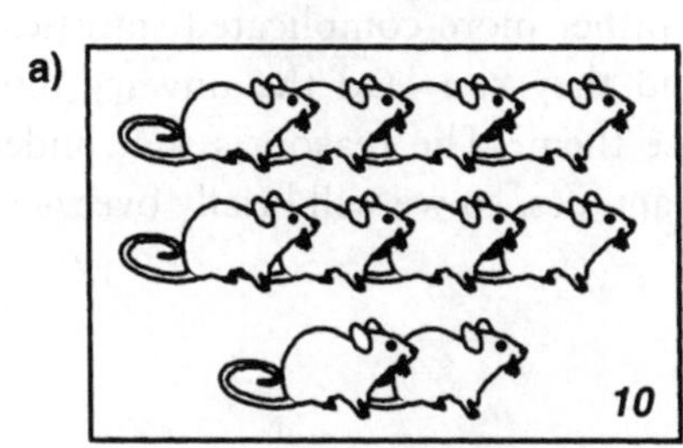

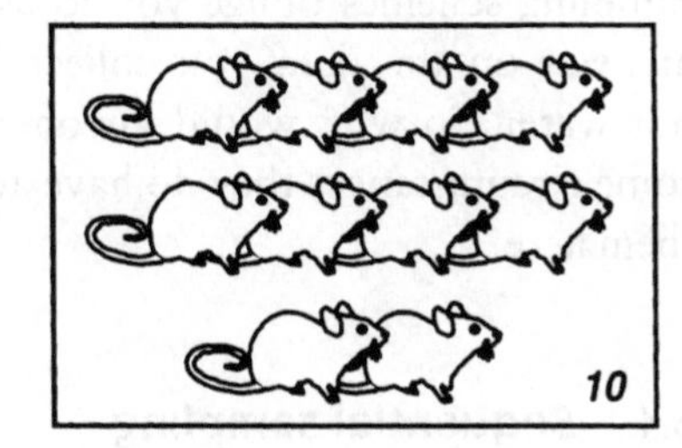

Most powerful design with 20 mice

b)

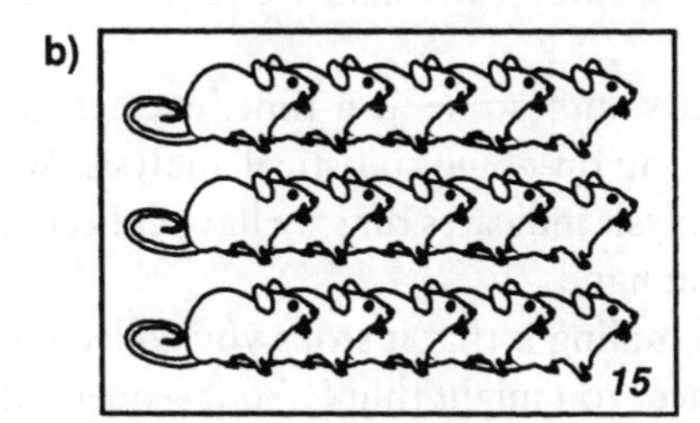

Less powerful design using 20 mice but fewer mice are subjected to the experimental treatment

c)

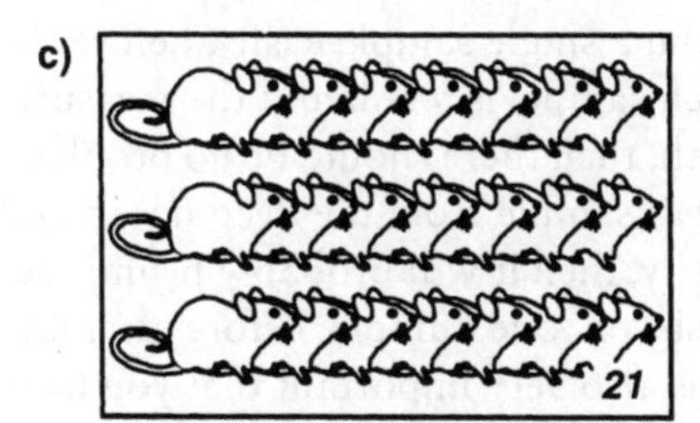

More mice are used overall but the statistical power is better than (b) without subjecting any more mice to the experimental treatment

Figure 6.3 Three of the possible alternative combinations of group sizes for an experiment on mice.

You should almost always try to use balanced designs—but keep in mind that it may sometimes be worth paying the cost of an unbalanced design to reduce the numbers in a particular treatment group.

6.4 Other sampling schemes

In Chapter 3, we emphasized the use of random sampling as a way of avoiding many possible problems of experimental design such as confounding variables and pseudoreplication. However, there are some other ways to sample, other than completely randomly. In general, these sampling schemes oblige you to use rather more complicated statistical analysis on the data you collect, and they can lead the unwary into hot water. So why would anyone use them? The reason is that under some circumstances they do have advantages, so we will briefly overview them here.

6.4.1 Sequential sampling

Previously we have considered cases where we decide on how many samples to take, collect these samples, extract the data we require from them, then do the analysis.

In sequential sampling, we collect samples one at a time, extract the data from each and add those data to an on-going statistical analysis. We stop sampling when the on-going analysis indicates that we have collected enough data to answer the question at hand.

The big advantage of sequential sampling is that it stops you collecting more samples than you need to. Hence, you might think about sequential sampling if sampling is costly, or where there are conservation or ethical implications of excessive sampling.

However, you can see that sequential sampling is only practical if measurements can be extracted from a single sample easily, before the next sample is taken. Hence, if each sample is a fish and the measurement you require from it is its length, then there should be no problem. However, if the sample is a sediment sample from the deep ocean and the measurement is microbial density, then it will probably be impractical to do the laboratory analysis on one sample before deciding whether or not to take another. It is also very important that you have an objective stopping criterion defined before you start sampling, so that no-one can accuse you of collecting data until you had the answer that you wanted, then stopping. For this reason alone, you should consult a statistician before attempting sequential sampling. However, as background reading for your meeting with the statistician, try Krebs (1999).

6.4.2 Stratified sampling

Imagine that you wanted to estimate the numbers of moths in a valley. The valley is made up of open fields and wooded areas. In a completely random sample, you would place traps randomly throughout the valley, and there is nothing wrong with this. But say you know that the moths are much more common in woods than open fields, but woods are relatively uncommon in the valley. If you distribute traps randomly, then by chance you may not select any wooded sites. In such cases you can sample more effectively by stratifying. To achieve this in our example, you divide the sampling area into the woods and the fields, then randomly allocate traps within each of these two strata (see Figure 6.4). Although the statistics will be slightly more complex, this will allow you to get a more accurate answer for the same effort as totally random sampling. The reason for this is exactly analogous to the motivation for blocking in section 4.3: we are partitioning variation. You are not limited to two strata, and you need not only stratify on habitat type. In a survey on links between gender and food preference, you might choose to stratify on age or socio-economic group or both. However, be warned: the more strata you have, the more complex the final statistical calculations become. Of course, once we have our estimates of the densities of moths in the different strata, we still don't have an estimate of the moth density in the whole valley. To make such an estimate we will also need to know relative frequencies of strata within the whole sample, in our case the relative frequency of wooded and open land in the valley. This is never as straightforward to measure as it seems. So stratifying can be very effective, but you need to think carefully about how to apply it to your sampling, and you need to think about how it will affect the ease of analysis of your data. Again, Krebs (1999) provides a gentle introduction to this technique.

6.4.3 Systematic sampling

Let's say that you want to get a sample of patients visiting their doctor to fill out a questionnaire. Our advice up until now is that you should ask the doctor to give the questionnaire out to patients that were selected randomly from the list of patients visiting them. This will take some organizing, so your temptation is to simply ask the doctor to give it to every 10th patient, say. Is there anything wrong with this systematic sampling? One line of thought is that there is nothing wrong with this because there is nothing systematic about the order in which doctors see patients, and so a sample collected this way could be treated exactly as if it were a random sample. We agree with this, but would caution you. First, any statistical analysis you do will assume that you sampled randomly. A systematic sample is NOT a random sample, so you are giving Devil's advocates a possible way to criticize you. Hence only think about systematic sampling if the organizational saving over random sampling is considerable. After all, how much effort would it really be to generate a list of random numbers and use that to select the patients before they arrive? Secondly, before sampling systematically,

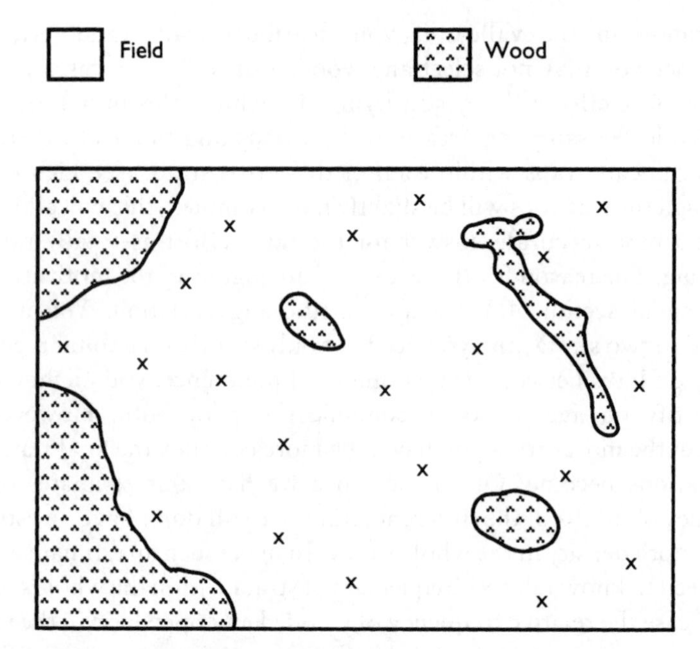

20 traps distributed randomly. (Notice that by chance none ended up in a wood, which makes up 10% of the area.)

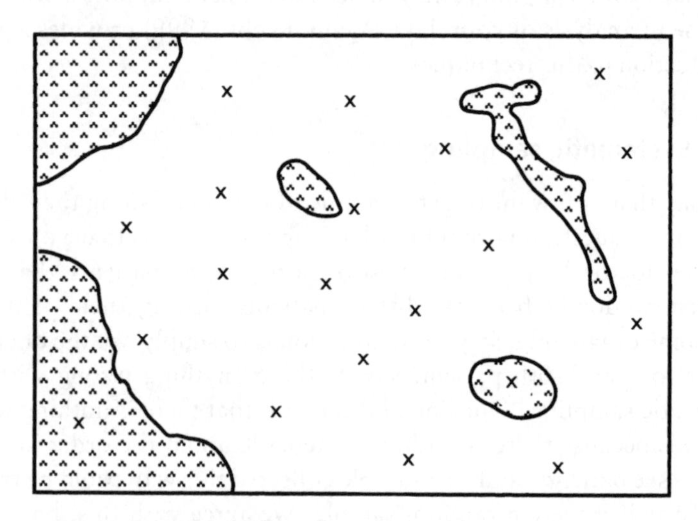

Now we stratify, assigning 10% (2) of the traps randomly to the woods and 90% (18) to the fields.

Figure 6.4 Comparison between random and stratified sampling.

sit down and think very carefully to make sure that there is no regularity in the data that would make systematic samples unrepresentative. Try to think about what the Devil's advocate would say in this situation: what pattern would they see that you have missed? To see how systematic sampling could lead to problems, see the example of report writing in section 4.6.

Non-random sampling techniques can be effective, but they need very careful handling.

6.5 Latin square designs

The Latin square design is a simple extension of the randomized complete block design that involves blocking on two characteristics. These designs are highly constrained in that the number of levels of each of the factors must be the same. For example, imagine we are interested in comparing the effects of three different diets on the growth rates of puppies. We are concerned that the breed of a puppy and the size of the litter it came from might be confounding factors. Now because diet has three levels (the three separate diets), then to use a Latin square design we need to impose that the other two factors also have three levels, so we would use three different breeds of dog and select individuals from three different litter sizes. This is quite a restriction. Let's say that to get the number of replicates you need, you have to use seven different breeds of dog. You could get around this problem, if you could divide these seven breeds into three groups. We would only recommend this trick if there is a truly natural division (e.g. into gun dogs, lap dogs and racing dogs, in our example). If you cannot achieve this symmetry, then you must use a fully randomized factorial design instead (see section 4.2). An important assumption of the Latin square design is that there are no interactions between the three factors. In our example, we are assuming that the effects of breed, litter size and diet act completely independently on the growth rates of puppies. Our knowledge of animal physiology suggests that this may well not be true. Hence, again we would recommend a factorial design instead, where interactions can be tested. However, in cases where you can justify this assumption (of no interactions) from your prior knowledge of the underlying biology, the Latin square does present a very efficient design, in terms of reducing the numbers of replicates needed. In practice, we have found that you almost never find yourself in a situation where you can justify use of this design. However, if you want to prove us wrong, then some designs for three and four levels are presented in Figure 6.5.

We have found little practical use for Latin squares.

Some examples of Latin Square designs

3 x 3

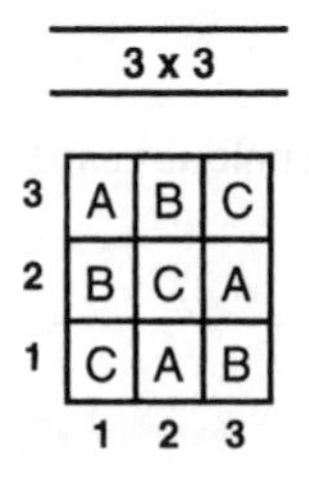

4 x 4

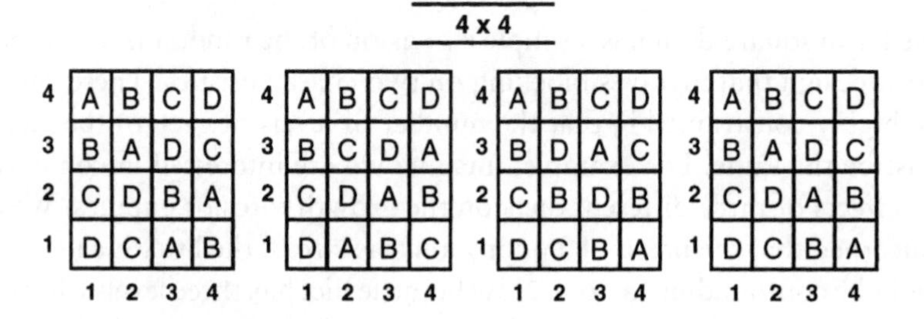

The first blocking factor (A) is represented by the columns and the second blocking factor (B) by the rows of these tables.

The treatment levels of factor C are represented by capital letters in the tables. Note the symmetry of these designs: each row is a complete block and each column is a complete block.

Figure 6.5 Some examples of Latin square designs.

6.6 More on interactions

6.6.1 Covariates can interact too

In Chapter 4 we discussed interactions where the effect of one factor depends on the level of another factor. However, another important type of interaction that you will come across is between factors and continuous variables (covariates). An interaction between a factor and a covariate means that the effect of the factor depends on the level of the covariate. Figure 6.6 shows situations with and without such interactions. Both panels show the results of experiments to look at the effect of the brightness of a male stickleback's mating coloration on his mating success under normal lighting and under red lighting. In Figure 6.6(a), there is no interaction. Mating success is higher in normal light, but the lines are parallel, and this indicates that the effect of light type is the same at all levels of male coloration. Contrast this with Figure 6.6(b). Here we see that the relationship between mating success and brightness differs under the two lighting regimes, i.e. there is an interaction. The effect of light type

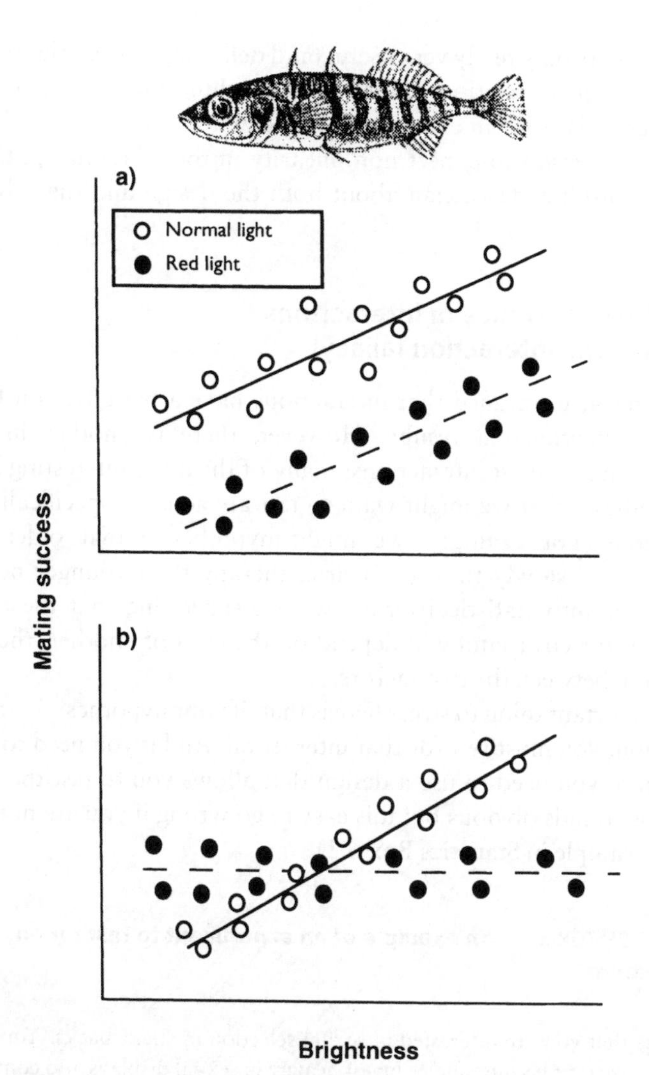

Figure 6.6 Two alternatives for the relationship between brightness of coloration and mating success for fish under two light regimes.

now depends on the brightness of the fish concerned, with normal light increasing mating success of bright fish, but having no effect (or perhaps a slightly negative effect) on dull fish. Hence, you should be aware that it is possible to design experiments that allow you to test for interactions between a covariate and a factor as well as between two factors. In general, this will require that the values of the covariate overlap to some extent for the different levels of the factor. So we would want to avoid an experiment where all the fish tested in normal light had high levels of coloration, but all those tested in red light have low levels of coloration. Similarly, as we have already mentioned in section 4.3.5 when discussing the use of covariates to control for differences between individuals, such techniques are only really very useful (and definitely only straightforward) if there is a linear relationship between the thing that we are measuring and the covariate: in our example, between mating success and coloration. If you have reason to expect non-linearity in the relationship, then you need to consult a statistician about both the design and the subsequent analysis.

6.6.2 The importance of interactions (and the interaction fallacy)

In Chapter 4, we argued that interactions have a big effect on how we interpret experimental results. However, there is another important reason to think about interactions: many of the more interesting biological hypotheses that we might want to test are actually specifically about interactions. For example, we might hypothesize that older people respond more slowly to a given drug therapy than younger people. If we put this into statistical terms, we are suggesting that the effect of one factor (or covariate) will depend on the level of another; there is an interaction between the two factors.

The important thing to stress here is that, if your hypothesis is about the interaction, you must test for that interaction. And if you need to test the interaction, you need to use a design that allows you to test the interaction. This sounds obvious but it is easy to go wrong if you are not careful (see the example in Statistics Box 6.1).

STATISTICS BOX 6.1 **An example of an experiment to test for an interaction**

Suppose that you are interested in sexual selection in sticklebacks. You might hypothesize that because males invest heavily in sexual displays and competing for females, they may be less good than females at coping with parasitic infections. One possible prediction of this is that any deleterious effects of parasites will be greater on male sticklebacks than on females. This is a hypothesis about an interaction. We are saying that we think that the effect of parasitism will depend on the sex of the fish, i.e. the effect of the first factor (parasitism) will depend on the level of the second (sex). By now, you will immediately see that one way to test this hypothesis would be to have a two-factor design, with infection status (parasites or no parasites) as one factor, and sex (male or female) as the other.

You might then measure the weight change in fish during the experiment as a measure of the cost of parasitism. As an aside, you might be wondering why you need the fish with no infections; wouldn't it be enough to simply have infected males and females and compare their change in weight in a one-factor design? The problem with this is that we would have no control. If we found that fish changed in weight, we wouldn't know whether this was due to the parasites or

STATISTICS BOX 6.1 (*Continued*)

something else. So, to measure the effect of parasitism, we need to compare fish with and without parasites. Now if we did the experiment this way, it would be a relatively simple matter to test for the interaction between sex and infections status, and if the interaction was significant, this would support our prediction (but remember that a significant interaction could also arise if parasites had a bigger effect on females than males, so it is essential to look at plots of the data to make sure that the trend is the way round that you predicted).

Now this (hopefully) seems straightforward to you now, but we claimed earlier that people often make mistakes, so what do they do wrong? The most common mistake is to treat this experiment as if it were two single-factor experiments done on the males and females separately. The researcher might analyse the difference between males with and without parasites, and find that males with parasites lose significantly more weight. The analysis could then be repeated on the females, and may perhaps find no significant difference between the weights of the two types of female. From this the researcher would then conclude that there is a difference in the effect of parasites on males and females.

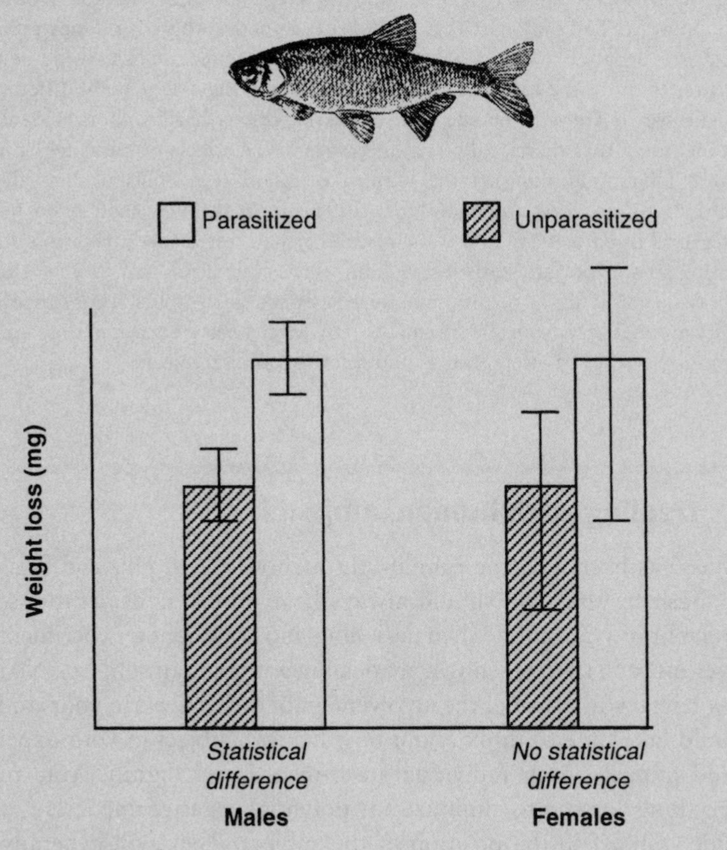

Figure 6.7 Results of an experiment exploring the consequences of gender and parasitism on weight loss in fish.

Q 6.2 An eminent professor hypothesized that plants from old tin mine sites will have evolved an increased ability to deal with tin poisoning. In a laboratory study, plants collected from mine sites grew bigger than plants from non-mine sites when grown in soil containing high tin levels. From this, the professor concluded that his hypothesis was correct. A bright young research student questioned this conclusion on the grounds that there was no evidence the increased growth rate had anything to do with tin; maybe the mine plants had evolved to grow bigger for some other reason. The professor repeated his experiment using tin-free soil and found no difference in the size of plants from the two groups. The professor now concluded that his hypothesis was certainly correct. Do you agree?

STATISTICS BOX 6.1 (***Continued***)

This seems straightforward, but it is wrong. The reason it is wrong is that the researcher has only tested for the main effects of parasitism, but not the interaction between parasitism and sex. By doing separate tests, the researcher has looked at the difference in male fish caused by parasites, and the difference in female fish caused by parasites, but has not directly compared the difference in male fish to the difference in female fish. You may be thinking that we are being needlessly fussy, but in Figure 6.7 we show how this may make a difference. Figure 6.7 shows the results of an experiment like the one that we have just described. You can see that the weight loss of male fish is greater when they are parasitized than when they are not, and a statistical comparison of these means would show a significant difference between them. In contrast, for the females, the statistical test says that there is no significant difference between the parasitized and unparasitized individuals. However, if we look at the Figure, we can see that the means values for the females are exactly the same as those for the males. What is going on? Why is the male difference statistically significant, but the female difference not? Well, there are several possible reasons, but the most obvious is that maybe the sample sizes for males and females are different. Suppose we had only half as many females as males; this would mean that our statistical power to detect a difference between the female fish would be much lower than for the male fish, leading to us observing a significant difference in male fish but not in female fish. However, this is hardly good evidence for concluding that there really is a difference in the effects of parasites on males and females. Looking at the graphs, common sense tells us that there is absolutely no good evidence for a difference in the way males and females respond to parasitism, and if the researcher had tested the interaction directly using an appropriate statistical test, this is the conclusion that they would have drawn. This is not just a problem caused by unequal sample sizes; it can arise for any number of reasons. In the end the only way of being confident that you have (or don't have) an interaction is to test for that interaction.

6.7 Dealing with human subjects

Ethics has been a theme running throughout this book, and it is worth emphasizing that you should always have ethical considerations at the forefront of your mind when designing and conducting experiments, and never more so than when you are dealing with human subjects. No matter how trivial you consider the involvement of the subjects in your study, you should take time to think about how being a subject in your experiment could impact on an individual's self-regard and dignity. You must do everything you can to minimize any potential negative impacts.

Put yourself in the position of the subject. They will generally be in unfamiliar surroundings, being asked to do things that they do not normally do. As the person instructing them, you are likely to be perceived as an authority figure. It is in your power to make the subjects feel as comfortable and relaxed as possible. It is very easy for subjects to feel that your data collection is a test in which their self-respect makes them want to perform as well as possible. Although they are just an anonymous 20-year-old female to your study, they can feel as if they personally are being tested. It is very easy for them to feel deflated after feeling that they performed less than perfectly. It's important that you do all that you can to avoid this feeling.

One important issue with humans is the idea of informed consent: that not only does the person agree to being part of your study, but they give their consent in the knowledge of why these data are valuable and how they are going to be used. As much as is possible, the subject should have forewarning of what is going to happen to them and what will be asked of them. Volunteers should also be fully aware that they are fully entitled to stop participating at any point and withdraw from the study without having to justify this decision to anyone. This may seem like a nuisance to you, and it is, but you must remember that high levels of drop-outs of this kind are likely to be caused by a failure to give proper forewarning of demands made on the subjects in your initial briefing of volunteers while you were collecting their informed consent. You will of course get the odd person who withdraws half way through the data collection for reasons that are impossible for you to imagine or understand. Don't stress about this; if you have designed your data collection properly, then the loss of this individual should not jeopardize your experiment, and you should take heart that you have created the appropriate atmosphere where people are comfortable acting on their right to withdraw rather than suffering in silence. Similarly, you will get the odd person who does not seem willing or able to co-operate in data collection as you expected. If this happens frequently, then there is something wrong with your data collection techniques, but if it is just a very rare individual, then don't get frustrated, just chalk it up as an interesting experience that reminds you of the tremendous variety of human personality. No matter what you think, treat the person with dignity and professionalism, collect the data from them as best you can, send them on their way without any feeling that you've been inconvenienced, and think at the end of the day (when you can look back on the incident with a broader perspective) whether you should discard that individual from your data set or not.

Q 6.3 On what grounds might you delete data from an individual from your study?

No matter how fully people were briefed beforehand, you should always make sure that your experimental schedule does not have the subjects bundled back out onto the street within 15 seconds of you collecting the data from them. Someone friendly and knowledgeable should be made available to them afterwards to answer any questions or concerns that arose during the data collection. Not all subjects will want to avail themselves of this opportunity, but it should be available to those who do. There will often be aspects to the work that you have to hide from the subjects before the data collection so as not to influence their performance or answers to questions, but that information should be made available to those who are interested after you have collected their data. This brings us on to the important concept of deception.

6.7.1 Deception

Sometimes the logical way to get the most accurate information from a subject would be to deceive them. Imagine that you want to simply measure the prevalence of left-handed writing among people in the local population. You feel that the best way to do this is to ask volunteers to sit at a table and fill out a short form with a pen, both of which are placed centrally on the table directly in front of where the subject is asked to sit. You then note down whether the person uses their right or left hand. You are collecting data by deception. Is this ethical? Scientists have different standpoints on this. Some would say that there are no circumstances where collecting information by deception is appropriate, and this is a viewpoint that we consider to be entirely valid. Our viewpoint is that there are some circumstances where addressing a valid scientific question can only be practically addressed by collecting the data by deception. If you can demonstrate both that your question is worth asking, and that it can only be addressed by deception, then such deception may be appropriate. Returning to our example, we would consider that measuring the local prevalence of left-handed writing could well be a scientifically valid thing to measure, but we are not convinced that these data could only be collected by deception. Firstly, we are not convinced that simply asking people whether they write with their right or left hand (or both) would not be effective. It seems unlikely that people would be unaware of this, and we can see little motivation to be dishonest in their answer. However, if you were concerned about dishonesty, then perhaps you could collect data on buses and trains by watching people fill in crossword puzzles in their newspapers. Thus, in this case (and the vast majority of cases) deception cannot be justified. Even if deception can be justified, we would strongly recommend that in almost all circumstances you then de-brief your subjects, explaining the nature of the deception, the purpose of the data collection, and why you felt that deception was necessary to collect the data. You should then ask them if they are comfortable with the data being used for the stated purpose, and explain that you'd be entirely prepared to strike their data from your study if they feel at all uncomfortable with your use of deception. If they do feel such discomfort, then you should apologize, explaining your motive again if appropriate.

6.7.2 Collecting data without permission

In the last section we suggested that you could collect data on handedness by observing people filling out crossword puzzles in their newspapers in public places like buses and trains. Is this ethical, in that you are collecting data from people without their knowledge? We would be comfortable with data collected in this way provided that the data contains nothing that would allow specific individuals to be identified if someone were to get hold of your notes, and providing that the information is in the public domain. By this we mean that their behaviour or other measure is something that the subject is comfortable with other people observing. If you fill out a crossword on a train, then you can be presumed to be unconcerned about allowing others to observe you in this act. Thus collecting data on aspects of this should not be considered intrusive. But looking through people's dustbins to see what foodstuffs they eat or peering into kitchen windows to record what time of day meals are prepared are unduly obtrusive.

You should also take care that data collected in this way does not involve you in anything that would make people feel uncomfortable. Say you want to know how many shops the average shopper visits in half an hour. Picking people at random and following them around your local shopping centre for 30 minutes would be irresponsible and unethical. Imagine how you would feel if you noticed a complete stranger tailing you around the shops. We'd phone the police in such a situation.

6.7.3 Confidentiality

There are very few experiments where it is necessary to record someone's name. If you don't need their name then don't record it. There may be circumstances where you do need to record personal information such as names and telephone numbers, perhaps if your study involves repeated follow-ups. You must be very careful with those data, never allowing it to fall into other hands. It would also be wise to avoid keeping these data in the same place as sensitive information about individuals. For example, if you require individuals to fill in personality questionnaires at 6-monthly intervals, you should give every subject a code number (or pseudonym), and this code rather than their name should be used in your files that contain the personality information. Have a separate file, kept quite apart from the personality information, that holds the name and contact details of individuals along with their code numbers. This may seem like unnecessary hassle for you, but think how you'd feel if someone else's carelessness allowed your details to fall into the wrong hands. So treat confidentiality very seriously, and take time to explain to your subjects how seriously you take it.

6.7.4 Discretion

Sometimes you must obtain sensitive information from people, and you must go about this in a sensitive way. We have no problem with asking people in the street whether they are left or right handed, as this question does not require a lot of thought to answer and is not likely to cause discomfort to the person questioned. However, asking someone about the number of sexual partners they have had in their lifetime, or their views on the ethics of eating meat, does not fall into this category. Hence, these are not appropriate questions to ask someone after stopping them in the street. You will get more accurate data and protect your subjects better if you ask them these questions in a more appropriate environment. For example, you might still stop people in the street and explain that you are surveying people's opinions on the ethical implications of what we eat, asking them if they would be comfortable filling out a questionnaire for you in their own time and posting it back to you in the freepost envelope provided. Or you might ask them if they would take away a flier that explains the purpose of the survey and invites those interested in taking part to phone a certain number to arrange a time when they are willing to come to your institute to be interviewed. Now each of these solutions has problems of introduced bias, because there is the potential that a non-random sample of people asked will respond (see our comments on self-selection in section 3.4.3), but by careful design of your questionnaire or interview, you can evaluate the extent of this bias. For example, you could collect data on the age of respondents and compare this to existing data on the age-structure of your statistical population to explore whether young people, for example, seem more willing to participate. You can then combat any highlighted deficiencies in your sample by targeting certain types of people in subsequent recruiting sessions. In contrast to this, asking people questions on complex or sensitive issues in the street is certain to produce poor-quality data and give you no real way to even valuate how poor they are.

6.7.5 Ethical guidelines

The institution under whose auspices you are carrying out experiments is almost certain to have a set of ethical guidelines and/or an ethics committee that has to approve experiments that involve humans. If you are carrying out the experiment then it is your responsibility to make sure that you are complying with the local ethical requirements. Do so in plenty of time before the start of the experiment, in case your particular experiment needs approval by a local ethical committee, since such committees may sit only infrequently. If your institution does have ethical guidelines, then it almost certainly has sanctions to impose on those that do not comply.

If in doubt, you should, of course, seek advice from others about the institution's policies. But make sure you get authoritative advice. If you are carrying out the experiment then the onus is on you to have the correct permissions.

If you are planning to publish your work then journals too often have published sets of ethical guidelines. It is worth being familiar with those of likely target journals for your work.

6.7.6 Volunteers

Volunteers are great! It is amazing what good-hearted people are willing to put up with (often for no financial reward) to help someone out with a scientific study. Science would be in a lot of trouble without volunteers. But take time to think what population you are looking to sample from and whether your volunteers are an unbiased sample from that population. Imagine that you are stopping people in the street and asking them whether they are right or left handed. If the population that you are interested in are adults in your local area, then you would doubtless be careful to collect samples from different streets and at different times of day so as to get as representative a sample as possible. You'd also be careful to make sure that you are unbiased in your approaches to people. This can easily occur, as it is perfectly normal to feel more comfortable approaching people of your own age group, for example, in such situations. Even if you avoid all these potential sources of bias in your sampling, there is always the danger of bias occurring because not all people approached will volunteer to answer your question. If, for example, handedness was more strongly associated with personality traits like openness and helpfulness, then your data set could be biased because one group (right or left handers) would be more likely to participate in your study than the other. You should weigh up the likelihood of such biases being a problem in any study using volunteers, and if there is a reasonable likelihood of such a problem in your study, then redesign your experiment to avoid this problem or to allow you to evaluate whether such a bias was a significant factor in your data collection.

6.7.7 Honesty of subjects

When your data collection involves asking people questions, you have got to consider the possibility that people may be dishonest in their answer. This can occur for a number of reasons. It may be that they are trying to give you the answer that they think you want to hear. It could be that they are seeking to impress you, or that they find the honest answer embarrassing. There are a number of things you can do to combat this. Most importantly, you can create an atmosphere in which the subject understands the need for honesty, and feels that your consideration of confidentiality and discretion will avoid any need to lie for reasons of potential embarrassment. This atmosphere is created by your initial discussions when you recruit the subjects and obtain informed consent. It is not just about what you say, but how it is said and by whom. Maintain a friendly but professional atmosphere. It is best if the questioner is seen to be disinterested in the subject except as the subject of a scientific study. That is, if you have some relationship with the answerer (e.g. you are a colleague, or a health care professional involved in the subject's treatment), then it's probably better if you get someone else to conduct the interview. Explicitly warn subjects about giving the answers that they think you want to hear, and aid this process by avoiding asking leading questions. Lastly, if you are gathering information about issues where dishonesty is likely (e.g. illegal activity or sexual behaviour) then it may be wise to ask a series of questions that effectively collect the same information, allowing you to check for inconsistency in responses, as indicators of dishonesty. You might choose to (tactfully) confront the subject with this apparent inconsistency during the interview in the hope of getting to the truth, or delete such inconsistent individuals from your sample.

6.7.8 There is no perfect study: a reprise

In Chapter 2, we made the claim that all studies have limitations, and nowhere is this more likely to be the case than in studies involving humans. Sometimes, despite all your planning and best efforts, you will not be able to totally exclude the possibility that self-selection or dishonesty will affect your results. However, even in these situations, an awareness of the possible biases will allow you to consider how, and whether, they are likely to affect your conclusions. If your question is worth asking, and you have addressed it in the best way available, then no one can ask for more. Being clear about the limits of what can be concluded from a data set is as important as the conclusions themselves.

Ethical and practical complications can arise when the study species is our own—take very great care.

Summary

- When deciding on the levels of a treatment, if you expect a factor to have a linear effect, and all you are interested in is quantifying that linear effect, then the best design is simply to have two levels at the lowest and highest biologically realistic values and to divide your experimental subjects equally between these two treatment levels.

- Using only two levels can be misleading, however, if the factor has a non-linear effect. In this case, you should use a large number of treatment levels in order to explore the shape of the relationship.
- Sometimes you can improve the information on each independent experimental unit by independently measuring several subsamples from each unit.
- There may be a practical trade-off between the number of samples you can process and the number of subsamples you can take.
- You should aim to bias your design to use more samples if between-sample variation is likely to be higher than between-subsample variation. However, you should bias your design the other way if subsamples within an individual subject are expected to be very much more variable than between individual subjects. The first of these situations is much more common than the second.
- A balanced design gives maximum power for a fixed number of experimental subjects. However, the same power can sometimes be achieved by reducing the number of subjects given the experimental treatment and considerably increasing the number in the control group. Although this means that more subjects are used overall, it reduces the number subjected to the treatment. This is attractive if the treatment (but not the control) is particularly stressful, expensive or time consuming.
- Sequential, stratified or systematic sampling may on occasion be preferable alternatives to random sampling, but each must be used with caution.
- We introduce you to Latin square designs, but do not recommend their use without extremely careful thought.
- You can have interactions between a factor and a covariate as well as between two factors.
- If your hypothesis is about the interaction between factors (or between a factor and a covariate), then you must collect data that can be analysed to test for an interaction.
- Experiments with humans have their own ethical and practical challenges, so take extra care.

Sample answers to self-test questions

Remember that these answers are not the only possible answers to the relevant questions. Many of the questions are deliberately open-ended, and your answer might be quite different from ours but be equally as valid.

Chapter 1

Q 1.1

The question is rather vague as to the correct population to use. In particular, it mentions no geographical restriction. It is unlikely that the questioners are interested in all the prison inmates in the world. Hence, for convenience, we will assume that the questioners intend that we limit ourselves to the UK, but you could equally as validly assume any other geographical restriction. Since the prison population is also fluid, we also need a temporal restriction in order to exactly specify the population. Hence, if we decide to do our sampling on 12 March 2006, then the population from which we draw samples might most accurately be specified as 'all people being held in UK prisons on 12/3/06'. We should always look to be as specific as possible when describing our population, something that the original question did not achieve very well.

Q 1.2

For religious practices they are certainly not independent. Your expectation is that being locked in prison is a life-changing experience, where someone's openness to new religious ideas might be unusually high. Being locked in the same small room as someone who has strong religious feelings is highly likely to have a strong effect on someone's own religious feelings. Left-handedness is likely to be much less open to such social influences, so it might be reasonable to expect that the two cell-mates could be treated as independent samples for this part of the study. As with so much of experimental design, you'll see that arguments about statistical-sounding concepts like 'independence of data points' are addressed by our knowledge of the system rather than by mathematical theory.

Q 1.3

Natural selection tends to reduce variation, since more beneficial traits flourish at the expense of less beneficial ones. It is hard to see how the exact patterning on our fingers would have any effect on the probability of our genes being passed on to future generations. Indeed, a lack of correlation between fingerprint patterns in closely related individuals suggests that this patterning is not under tight genetic control. Hence, we have a lot of variation in this trait because it is not under strong selection pressure.

Q 1.4

Take your pick from hundreds! We'd certainly expect age, gender, genetics, occupation, diet and socio-economic group all to have the potential to affect eyesight. The problem is that these factors are also likely to influence someone's propensity to smoke, so teasing out the effects of smoking from other factors will be challenging.

Q 1.5

You could recruit a colleague to take measurements of one flock at the same agreed time as you are taking measurements on the other. However, this simply swaps one confounding factor (time of measurement) for another (person doing the measuring). Perhaps this would be an improvement if you could take steps to make sure that you both measure sheep in exactly the same way. Such steps could include both agreeing to a very explicit fixed protocol for sheep measurement and using the same type of equipment as each other. You could perhaps take half the measurements on one flock in the morning, then swap over with your colleague to take half the measurements on the other flock in the afternoon. You could then statistically test to see if the samples taken by the two different people from the same flock are significantly different from each other: if they are not, then this would give you confidence that you do not seem to have problems with identity of measurer or time of measurement as confounding factors in your study. However, there are lots of alternative ways you could have approached this study, this is by no means the only solution.

Chapter 2

Q 2.1

It may be that the people are (at least subconsciously) anxious to begin tackling the day's tasks at work as quickly as possible, whereas they are more relaxed about time on the way home. It could be that the penalty for arriving late at work is greater than the penalty for arriving late home from work. It could be that people like to lie in their bed as long as possible in the morning, then drive quickly to make up the time. It could be that the roads are generally busier when people return from work and so people are more likely to be delayed by slow-moving traffic on their return from work.

Q 2.2

It could be that regularly playing computer games does increase individuals' propensity to violence. One prediction of this is that if we found some volunteers who had not previously routinely played computer games and tested their propensity to perform violent acts both before and after a period of regular games playing, then their violence rating should increase.

Another hypothesis is that computer games do not influence propensity to violence, but people who have a high propensity to violence for other reasons (be they genetic, environmental or both) happen also to enjoy playing computer games. If this hypothesis is true, then we would expect that if we performed the experiment described above, then the volunteers' violence ratings would be unchanged by their period of regular games playing. Thus the experiment would allow us to test between our two rival hypotheses.

Q 2.3

Asking someone 'please keep a note of how often you commit violent acts over the next 6 months' or 'how often have you committed violent acts in the last 6 months?' is unlikely to produce very accurate answers. Firstly, people will disagree on what constitutes a violent act. Even if this can be overcome by agreeing a definition, then this is the sort of question about which you have to expect people to often be less than honest in their replies. We could stage situations involving potential conflict and monitor the reactions of our subjects. However, as well as being logistically very challenging, this is ethically highly questionable. Further, it is not clear to what extent someone's behaviour in one staged situation acts as a description of their general propensity towards violent behaviour.

Q 2.4

Firstly, can you find the specific pedestrian walkway and gain access to it? Can you see all the cars that are passing beneath you? Can you count all the cars in all the lanes that go in a particular direction simultaneously, or should you restrict yourself to one lane at a time? Can you count accurately using paper and pencil or should you use a tally counter? Can you count cars for 1 hour without a break without making mistakes, even in the wind and rain? Will you be distracted by other people crossing the bridge? Will you inconvenience other bridge users? Can you decide on an appropriate definition for a car? For example: is a Range Rover a car? Is a hearse a car? Do you count cars on a car transporter?

Q 2.5

Foraging behaviour is a plausible third variable. It is plausible that different badgers have different diets (perhaps through variation in local food availability) and that some food types are associated both with a higher risk of infection with gut parasites and with lower calorific value, leading to lower body weight.

Notice that if such a third-variable effect is operating, this does not mean that intestinal parasite load has no direct effect on body weight: both the third-variable and direct effects could be in operation, but further manipulative study would be required to estimate the relative importance of the two mechanisms for body weight.

Q 2.6

(a) There is a danger of third-variable effects here. For example, it seems likely that those who feel themselves to be time-stressed might (separately) have a high propensity to use highly processed foods and a high propensity for depression. There is also a possibility of reverse causation, where people who are depressed for whatever reason are less likely to be motivated to make meals from scratch and more likely to turn to easier-to-prepare highly processed foods. Manipulating someone's diet is reasonably straightforward, and results ought to be seen after weeks or months. So we would opt for manipulation.

(b) Again there are a lot of potential third variables here, e.g. sex, age, socio-economic group, ethnicity, other dietary factors (such as salt content), genetic factors and exercise. Hence, you are drawn towards manipulation. But heart attacks are relatively uncommon, and so even with thousands of people in the study, you would require the study to run for several years to hope to see any effect. There is also serious ethical concern about manipulation here, in that asking a volunteer to take a diet that you expect may increase their chance of a life-threatening event like a heart attack seems very hard to justify. In contrast, with careful monitoring, an increased risk of depression in (a)—while still an ethical issue—is perhaps more justifiable. These practical and ethical concerns would turn us away from manipulation in (b) and towards adopting a correlational approach, accepting that our data collection and statistical analysis will have to be carefully designed to control for several potentially important third factors.

Chapter 3

Q 3.1

Probably not. Married people are a subset of the population, and we would be safe to extrapolate from a sample of married people to the whole population only if we could argue that a person's likelihood of marriage and their height are likely to be entirely unrelated. However, this would be a difficult argument to make. Height is generally considered to be a relevant factor in people's attractiveness, which again might presumably be linked to their likelihood of marriage.

Q 3.2

We would be concerned that a given female's choice could be influenced by the decisions made by others, thus making individual choices no longer independent measures of preference. This might occur because individuals are more likely to adopt what appears to be the popular choice, simply because of a desire to be in a group with a large number of other (female) birds. Alternatively, we might find that competition or simple lack of space caused some individuals to move to the side of the cage that they would not have preferred if they had had a free choice.

Q 3.3

For hair colour you would probably be safe to count each individual as an independent data point. You might worry slightly that there may be some pseudoreplication due to relatedness if people who are particularly nervous about visiting the dentist or who are facing quite extensive procedures are accompanied by a family member, but this will be sufficiently uncommon that you may be safe to ignore it. However, pseudoreplication is a real problem for conversation. Firstly, if one person is talking then the strong likelihood is that they are talking to someone else! Hence people's decisions to talk are not independent. Further, if one conversation is struck up between two people, this is likely to lower other people's inhibitions about starting up another conversation. Hence people cannot be considered as independent replicates and you would be best to consider the 'situation' as a single replicate, and record total numbers of men and women present and the numbers of each that engage in conversation.

Q 3.4

You should be concerned. While you have 360 red blood cell counts, these only come from 60 different people. The six counts taken from each individual are not independent data points for examining the effects of the drug on red blood cell counts.

The simplest solution is to use your six data points to generate a single data point for each person. In this case, the mean red blood cell count would probably be appropriate. While more complicated statistical procedures would allow you to use all of these data points in your analysis, even these procedures would not allow you to use the 360 measures as independent data points.

Q 3.5

The answer to this question depends on the biology of the system. In particular it will depend on how much variation there will be between the three counts of the same sample, and also between the two samples from the same individual. If we know that there is little variation between counts of the same blood sample, then a single count of each sample might be sufficient. However, if there is a great deal of variation between smears, then taking a number of measures from a single sample and generating a mean value may give us a much better estimate of the 'true' cell count for the sample than any single count would. The same logic will apply to whether we need one or several blood samples from each individual. Thus, repeated sampling in this way may be an efficient way of dealing with some of the random variation in your system.

Q 3.6

An unfair question would be phrased in such a way to encourage the respondent (consciously or unconsciously) to answer in a particular way. Obviously unfairly phrased questions would include:

'Do you agree with me that Pepsi tastes better than Coke?'
'Recent surveys have shown that Coke tastes better than Pepsi, do you agree?'
'Sales of Pepsi continually outstrip Coke, do you think it tastes better?'
'I represent the Coca-Cola company, do you feel that our product tastes better than Pepsi?'

You might want to be careful with the very neutral 'Do you prefer Pepsi or Coke?' by alternating it with 'Do you prefer Coke to Pepsi?', just in case ordering has an effect on people's answers.

Q 3.7

Read the literature. Small mammal trapping is such a common scientific technique that a great deal is probably known about bait preference and 'trapability' in your study species. You would be particularly concerned for this study if there were reports of variation between the sexes in factors that might affect their probability of being trapped: for example general activity level, food preference or inquisitiveness. If these effects seem to be strong, then your survey may need a drastic re-think.

It might also be a good idea to vary potentially important factors, such as trap design, bait used and positions where the traps are set. This would let you explore statistically whether the estimate of sex ratio that you get depends on any of these factors. Strong differences in sex ratio obtained by traps with different baits, say, would suggest that sex differences in attraction to bait are likely to be a concern in your study.

Q 3.8

Yes! How do you know that the rodent that you find in a particular trap this morning is not the same individual that you pulled out of the trap yesterday morning? Two measurements of sex on the same individual would certainly be pseudoreplication. One way around this would be to mark individuals in a long-lasting but ethically acceptable way before releasing them; perhaps with a spot of paint on the top of the head. If you find any marked individuals in your traps, you can release them without adding them to your data-set.

Q 3.9

Your colleague is right to be concerned about animal welfare, but their blanket assertion of the maximum sample size should concern you. Would an experiment based on a small sample like this have any chance of detecting the effects that you are interested in? If not, then the birds would still suffer but with little hope of providing useful data. If you are convinced that the study is worth doing, then it may be better to use more birds and give yourself a reasonable chance of detecting the effect you are interested in. Otherwise you risk carrying out an uninformative study that then has to be repeated with a more appropriate sample size later on, and so the total number of birds used to address the question is actually higher than if you had done the experiment properly in the first place.

Chapter 4

Q 4.1

Samples in the control group should differ from the samples in the experimental group only in the variable or variables that are of interest to the experiment. This scientist has inadvertently added a confounding variable: time. Now any difference between control and experimental groups cannot be attributed with certainty to the variable of interest because it could simply be that conditions have changed over time. It is very difficult to persuade the Devil's advocate that the conditions, equipment or the scientist carrying out the procedures have not changed in some critical way between the time of the experimental subjects and the time of the controls. This problem is very similar to the problems of historical controls. Our colleague would have been well advised to avoid this situation by running experimental groups and controls concurrently and by randomizing the order in which procedures are applied to experimental units (see Chapter 5 for more on this).

Q 4.2

It is possible that the person observing the nest could be affected by an expectation of the effect of manipulation of clutch size. Concern about this could be avoided if they were blind to which nests received which treatments. Such a procedure requires that someone else carries out the manipulations. This would seem relatively easy to organize in most cases.

Q 4.3

It is tempting to assume that what this means is that the effect of one factor is affected by the values of the other factors, but actually the situation is more complicated than this. That would be the explanation if there were several interactions between just two of the factors. The interaction between three factors means something else again. The closest we can get to this is to say that not only does the effect of insecticide treatment depend on the level of food supplied (i.e. there is an interaction between pesticide and food), but that this interaction itself depends on the strain of tomato plant we are talking about. One way that this might happen is if food and insecticide affect growth rate independently in one of the strains, but in another strain adding insecticide has a much greater effect in high-food treatments than in low-food treatments. Put another way, it means that it is only sensible to look for interactions between two factors (e.g. food and pesticide) for each specific value of the other factor (each strain of plant separately).

Q 4.4

Possibly. Blocking is only a good idea if you know that your blocking variable actually explains a substantial fraction of variation: that is, the two scientists do really score differently from each other. We'd recommend a pilot study where each scientist scores independently a selection of tomatoes, and they then compare their scores to see if there is discrepancy between them. If there is discrepancy, they should try to understand their disagreements and work towards agreeing, then repeat the trial on another set of tomatoes to see if they are now in agreement. If they now agree perfectly, then we can stop worrying about variation due to the identity of the scorer. If they still cannot agree, then it is worth considering if it is possible for one person to do all the scoring and for them to agree on who that person should be. The moral of all this is that it is better to remove the source of the unwanted variation than to use blocking or other measures to control for it.

However, if both scientists are required and they cannot be relied upon to score the same tomato the same way, then blocking on identity of the scorer would be a way to control for this unwanted variation. Each scorer should sample half of the plants in each greenhouse and half of the plants in each treatment group. Be careful when allocating plants to scientists that you don't do this systematically by having a rule like 'Scientist A does the half of the greenhouse nearest the door'. If you do this and find an effect of scorer, you will not know if it is an effect of scorer or an effect of placement within greenhouses. Allocating plants randomly to scorers is going to be a non-trivial amount of work, so you may well want to re-think looking for ways to eliminate between-observer variation rather than controlling for it (see Section 5.4.2 for further discussion on this).

Q 4.5

The subjects are likely to be quite variable in their initial level of fitness, so blocking seems attractive. Although age and sex are easy to collect data on, we're not sure that we expect these to be strong contributors to between-subject variance in fitness. You'd be better off blocking according to some measure of initial fitness at the start of the experiment. You might ask volunteers to fill in a questionnaire about their existing level of exercise, or you might take measurements on them to calculate some index of body fat.

Q 4.6

The most important thing to establish is whether playing sounds to chickens has any effect on their laying; whether the nature of those sounds (classical music or conversation) has an effect is of secondary concern. Hence, we must make sure that we have a control that tests for the effect of playing sounds to the chickens. Therefore we need a control where we add the sound system (to control for any effect of the disturbance of adding the experimental equipment). Thus, a perfectly respectable design would be to have two groups (classical music and silent control) and randomly allocate 12 henhouses to each. However, the question is hinting that between-house variation might be expected to be low, so fewer samples in each group might be required; if we had another treatment, that would allow us eight in each group. Eight is an acceptable number, if we expect random variation to be low and drop-outs to be unlikely. Hence we feel that we can get away with having another control group to explore the question of whether (if there is an effect of sound) the type of sound matters. Here you have a lot of choice limited only by your imagination. It might not be unreasonable to select this further control with some consideration of producing an interesting story. It seems to us that 'hens prefer the Beatles to Beethoven' is a catchier headline than 'hens prefer classical music to the sound of aircraft landing' and so we'd be tempted to go for a second control group of pop music. Hence, we'd go for two controls: no sound and pop music as well as the planned treatment of classical music.

Q 4.7

Yes, although in this case they might better be termed deaf procedures! It would be a good idea for the eggs to be collected at a time of day when the experimental treatments are not being applied or by someone using ear plugs (providing the latter did not place the collector at any risk, e.g. from not hearing heavy machinery coming towards them). We'd especially want to put blind procedures in place when we think that our motivation for picking one control group in the experiment was for reasons of a sensational headline: that sort of thinking can easily spill over into subconsciously wanting one result more than another.

Q 4.8

It is likely that the farm keeps records of eggs collected from different henhouses. If so, it would be worth looking at these records to see if some houses consistently produce more eggs than others. If they do then it would be worth blocking on previous egg production.

Q 4.9

See section 4.4.2.

Q 4.10

Yes it would; however, this design will not be much (if any) easier or cheaper to perform than the counterbalanced one that we recommend, but is likely to be much less powerful. One reason for this is that we will have to make between-group comparisons, and the other reason is that we will now have only half the number of within-subject comparisons to detect an effect of the experimental treatment.

Q 4.11

You'll need to repeat the experiment several times. Our advice would be to run the experiment using all three incubators three separate times. At each time, each temperature is assigned to a different incubator, but the assignment changes between times such that each incubator is used to produce each of the three temperatures over the course of the three experimental times. You would then be able to test for any effects of incubator and time-of-experiment on growth rates in your statistical analysis. More usefully, your analysis will be able to control for any effects of these factors, making it easier for you to detect effects due to the factors that you are interested in (temperature and growth medium).

Chapter 5

Q 5.1

You should record measurement date for each chick and include it as a covariate in your statistical analysis. However, you may also want to explore whether the person doing the measuring is drifting over the course of the experiments: we suggest that you get another couple of properly trained measurers to take measurements of some of the same chicks as your main measurer: this should be done entirely independently of the measurements taken by the main measurer. Further, the main measurer should be told that some of their measurements will be checked independently but not which ones. Comparison between the three sets of measurements can then highlight any drift in the main measurer, since it is very unlikely that the three measurers will drift in the same way. Further, human nature being what it is, knowing that some of their measurements will be checked independently will make the main measurer particularly careful, making drift less likely in the first place. Finally, you should always bear in mind that, despite your best efforts, you may not be able to entirely remove or control for any effect of time in this study, and should consider what effects this might have on the results you have collected and the conclusions you can draw.

Q 5.2

In this case, we think not. Using two people with two stopwatches, some volunteers, and some care, you can easily collect data in a pilot study that should show that both imprecision and inaccuracy are not going to impact strongly on the results of your main experiment. Further, the cost of two stopwatches is going to be less than 1% of the cost of the automated system. It may involve increased labour costs, but labour is cheap or cost-free if you have friends. The set-up time of co-ordinating the stopwatch operators, practising working together and collecting the pilot data will probably be less than 10% of the set-up time of the automated system.

We might reach different conclusions if the times that were being recorded were much smaller, such that the relative importance of imprecision introduced by the stopwatches increased. Thus, if we were recording speeds of racing cars over 60 metres, then the extra hassle of the automated system would be worthwhile.

Q 5.3

We cannot take the approach recommended for the passport photos, because we cannot measure the same cats on more than one occasion. Even if cats could be brought in more than once, they may well have experienced weight change over the intervening time.

You must minimize the room for inadvertent variation in measurement technique. Have a strict written protocol for the measurement. This should include the posture of that cat while you are measuring the length of its back, with clear rules for identification of the two endpoints (top of the head and base of the tail). The mass should be taken on an electronic balance, calibrated at regular intervals. We expect greater scope for error in the length measurement than the mass measurement: hence we suggest that you measure the length at the start, then weigh the cat, before finally taking the length measurement again, giving you two estimates of the harder-to-measure variable on each cat.

We further suggest that you organize for a third party to repeat some of your measurements without your knowledge, in a similar manner to discussed for Q 5.1.

Lastly, if there are likely to be concerns about observer drift over time, then it is imperative that order of measurement is randomized across cats on different treatments if at all possible.

Q 5.4

You are introducing an unintended confounding variable: identity of observer. If you find a difference between two treatment groups you do not know if this can be explained in whole or in part by differences in the way the two observers recorded their data.

Q 5.5

Firstly consider whether you need to interview the subjects, or whether asking them to respond to written questions could be just as effective without the risk of observer effects.

Make sure that the interviewer is someone that the interviewees consider to be disinterested in them as a specific person. For example, they should be someone that the interviewee is unlikely to have encountered before or encounter again.

Train your interviewer very carefully not to (often subconsciously) encourage certain types of answers, and design your preparation of the subjects for interview so as to make the subjects less susceptible to observer effects.

More consideration of these issues is given in section 6.7.

Q 5.6

We don't know! But if your experiment involves collecting data in this way then you need to know! You need to do a pilot study where you collect data in this way for 30 minutes, spending the last 5 minutes re-scoring the same slides that you did in the first 5 minutes. If you have good agreement then it seems that fatigue is not a problem over 30 minute timescales, but without the pilot study you would not have known, and would have no defence against a Devil's advocate who expected that fatigue could easily kick in that quickly.

Q 5.7

Firstly, we think you need each subject to be evaluated independently by several psychiatrists who should not have a history of close association with each other. We would also recommend that if possible you avoid using psychiatrists who have previously expressed strong views on gender issues. And as in all situations where it is impossible to have the perfect design, if you cannot remove the potential for observer bias completely, you should be clear about what effect it might have on your results and conclusions.

Chapter 6

Q 6.1

Many aspects of the forests could affect beetle diversity. As well as your interest in tree type, variables such as altitude and land area covered could well have an effect. If we have only two forests, they are likely to differ in other ways as well as in tree type, for example differing in altitude. With only two forests we have no way of knowing whether any differences in beetle diversity found in the two sites are due to the variable of interest (tree type) or some third variable (for example, altitude). We need more samples so we can control for the effects of these potential third variables statistically.

Q 6.2

We would worry about this conclusion for a number of reasons. First, we would be concerned that the two experiments were carried out at different times, and so any differences between the experiments might be due to tin, but might be due to anything else that has changed in the lab between experiments. We would also want to know where the plants had come from, to reassure ourselves that they were really independent replicates. However, our main concern here is about interactions. The hypothesis is about an interaction. If the plants have evolved to be able to deal better with tin, we might predict that mine plants will grow better than other plants in the presence of tin, and the same or worse in the absence of tin. In other words, the effect of one factor (plant origin) depends on the level of the other factor (tin presence or absence). By carrying out two experiments and analysing them separately, the professor has not tested for an interaction, and so his conclusions are premature.

Q 6.3

(i) if you had an incomplete set of data on this individual and the missing values would make your statistical analysis more complex.

(ii) if the data were unusable—for example if you simply could not read someone's handwriting on a questionnaire form.

(iii) If you are confident that they are not representative of the statistical population that you are interested in. Be very cautious about deleting people on these grounds. 'Unrepresentative' does not mean 'unusual', 'unexpected' or 'inconvenient'.

Flow chart on experimental design

The following flow chart is intended to take you step by step through the different stages involved in designing an experiment. You can treat it a bit like a checklist that should be run through every time you plan an experiment to make sure you've considered all the important issues that we mention in this book. Taking account of these issues properly will make the difference between a successful experiment that's going to yield valuable results, and an experiment that might give results that are all but worthless. Think of it this way: the time you will need to spend designing an experiment carefully will only ever be a fraction of the time you actually spend gathering results. Just a few extra minutes early on should mean that you'll be able to feel much more confident during the many more minutes you're going to spend in the lab, or out in the field. It may seem laborious but, trust us, it will be time well spent. Good luck!

The numbers on the chart provide direction to the relevant sections of the book, for each part of the design process.

(i) Preliminaries

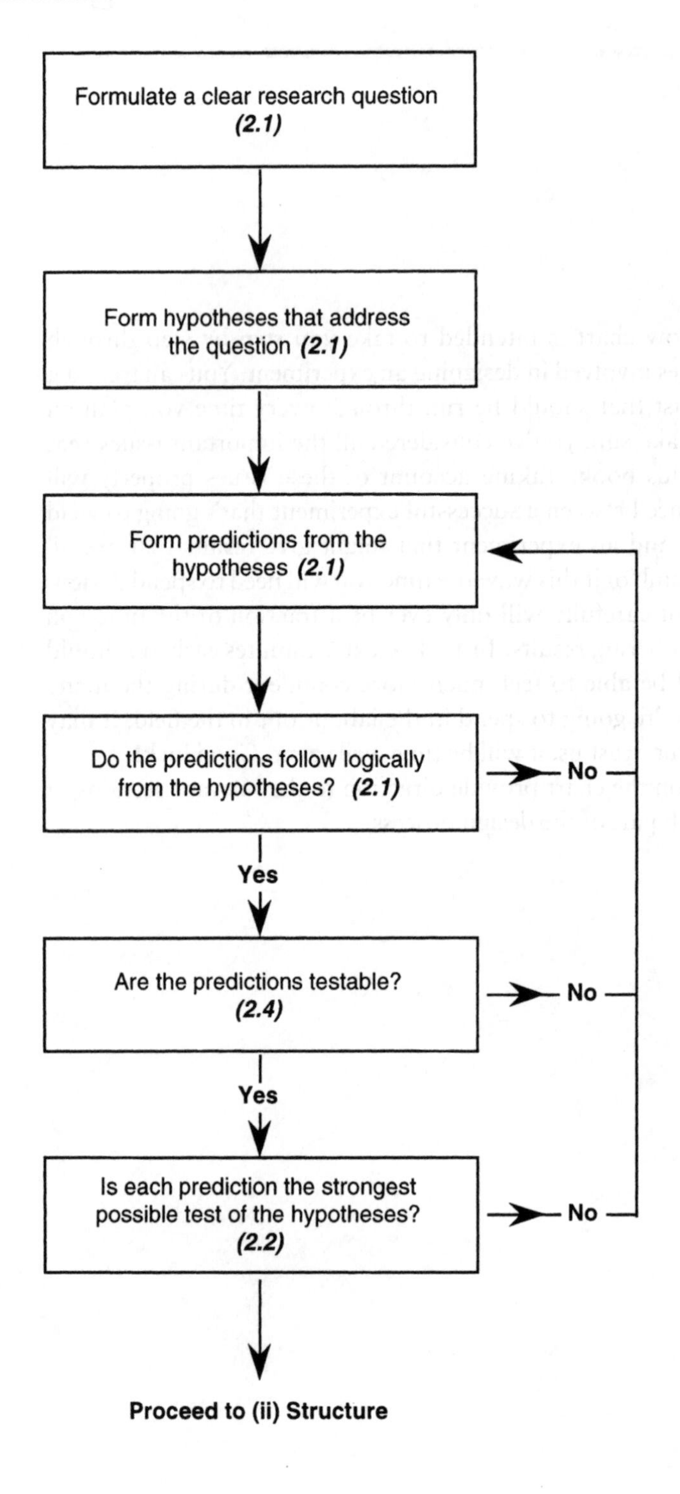

(ii) Structure

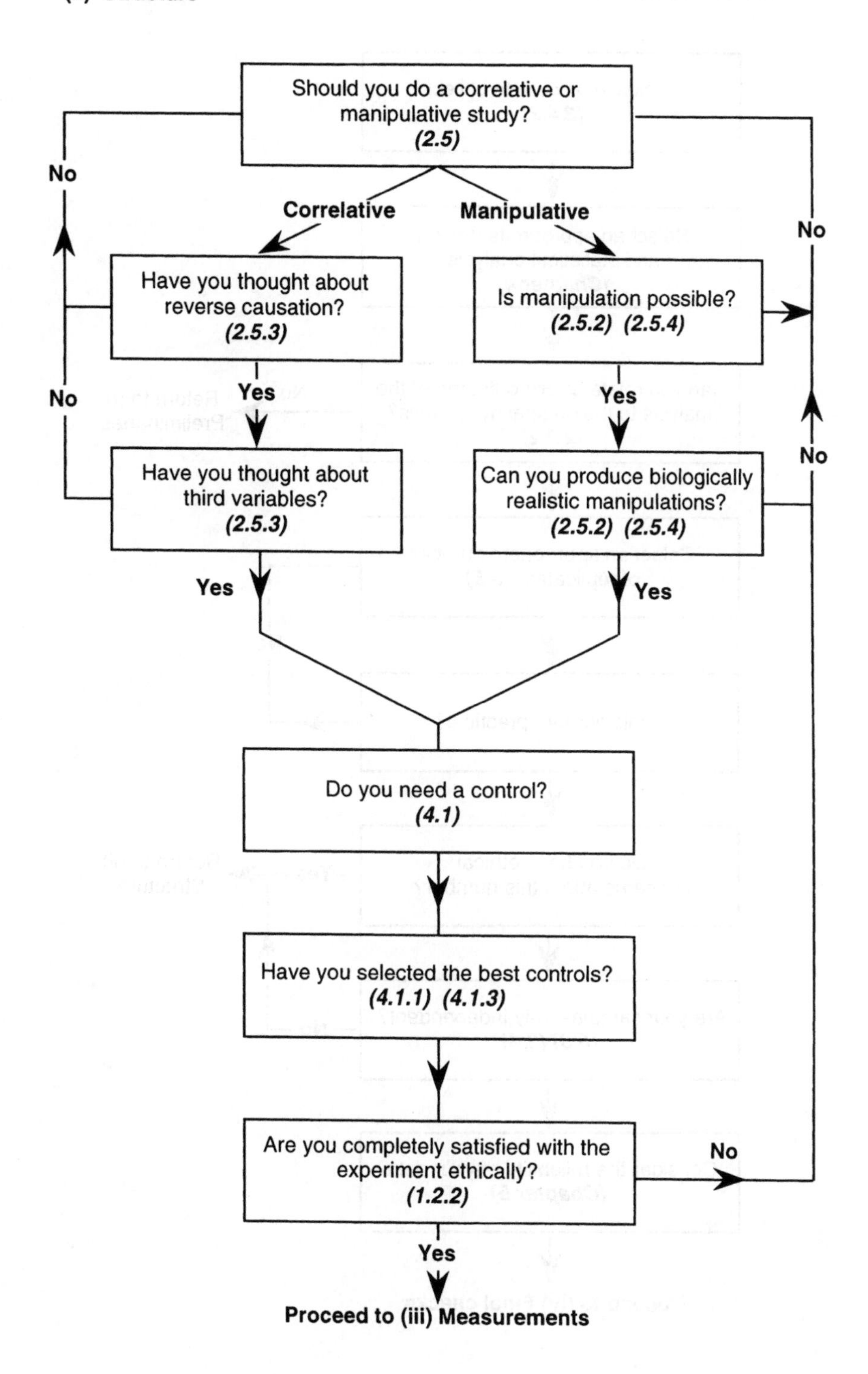

(iii) Measurements

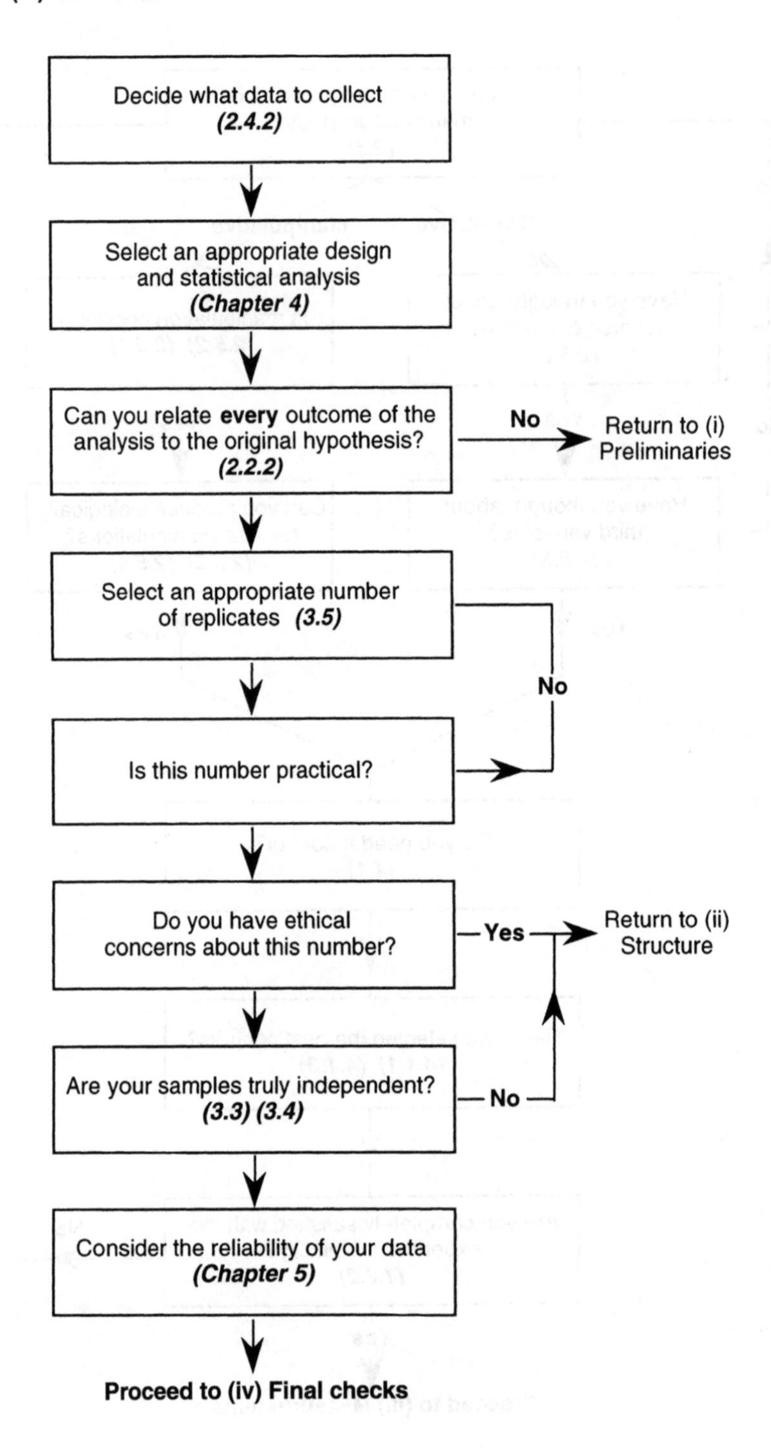

(iv) Final checks

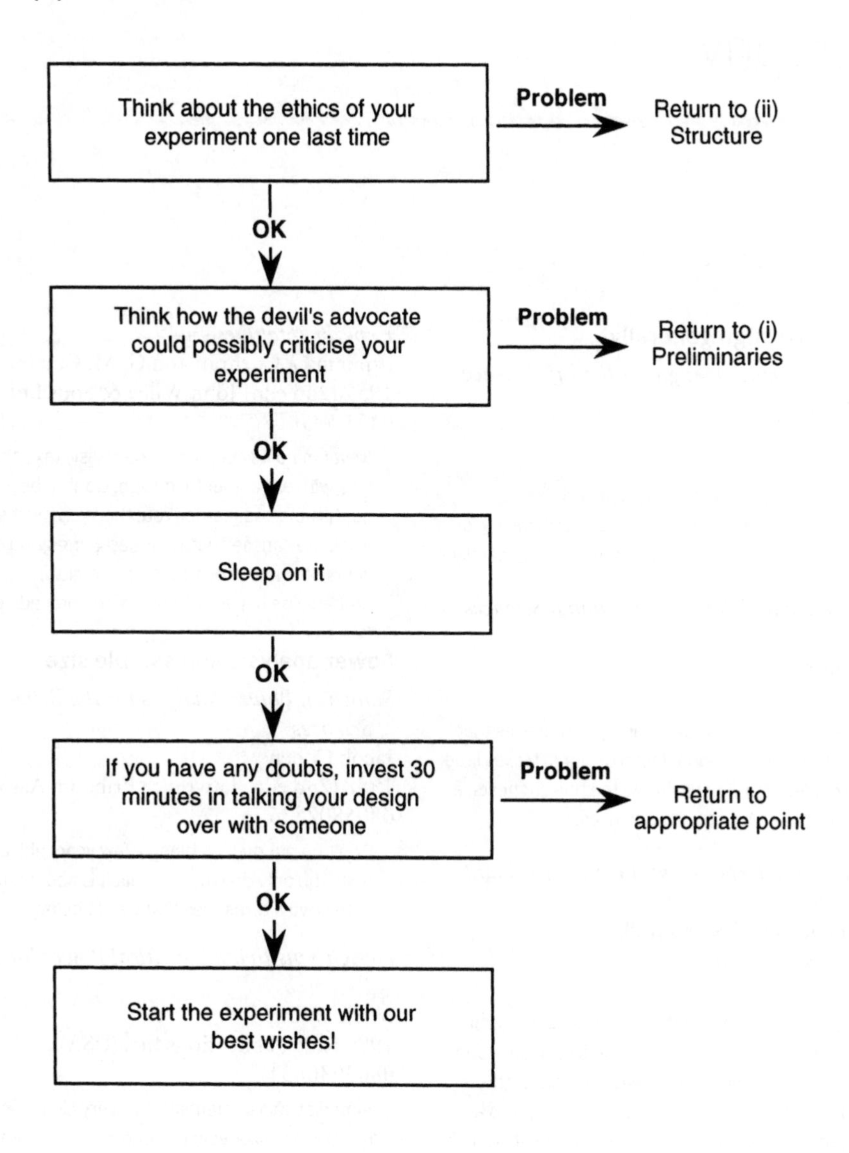

Bibliography

Further books on design generally

Principles of Experimental Design for the Life Sciences
Murray R. Selwyn
1996 CRC Press
0849394619

Very clearly written and very gentle in its demands on the reader's mathematical abilities. Recommended particularly for those interested in clinical trials and human medicine generally.

Experimental Design and Analysis in Animal Sciences
T. R. Morris
1999 CABI Publishing
0851993494

A gentle first step beyond our book, taking you to places that we don't go, such as towards unbalanced designs. Makes little demand of the reader mathematically. As the title suggests, a good buy for students of Agricultural Sciences.

Practical Statistics and Experimental Design for Plant and Crop Science
Alan G. Clewer and David Scarisbrick
2001 John Wiley & Sons Ltd
047899097

This is to be strongly recommended, whether you are a crop scientist or not. It is a very good introduction to statistics and links this very well with experimental design. It asks a little more of the reader in terms of mathematical ability, but is a model of clarity and interesting writing. Many scientists will never need to know more about experimental design and statistical analysis than is contained in this book.

Introduction to the Design and Analysis of Experiments
Geoffrey M. Clarke and Robert E. Kempson
1997 Arnold
9780340 645555

Compared to Clewer and Scarisbrick, this book draws on examples from a wider variety of branches of the life sciences, and demands significantly more from the reader in terms of mathematical ability. Otherwise it is rather similar.

Experimental Designs
William G. Cochran and G. M. Cox
1957 (2nd edn) John Wiley & Sons Ltd
0471 545678

Never mind its age, this is the classic text. If you have fallen in love with experimental design, want to become an expert, and maths holds no fear for you, then you must seek this book out. If you are satisfied with just designing good experiments yourself, without becoming an oracle for others, then you will probably be better with one of the preceding books.

Power analysis and sample size

Statistical Power Analysis for the Behavioural Sciences
Jacob Cohen
1988 (2nd edn) Lawrence Erlbaum Associates
0805802835

Don't be put off by it being a few years old; this is the definitive text. The only drawback is that it is about four times the price of the two alternatives that we list below.

Design Sensitivity: Statistical Power for Experimental Research
Mark W. Lipsey
1990 Sage Publications Inc. (USA)
0803930631

Aimed at social scientists, but very clearly written and very much in a similar vein to Cohen. So if you are unsure about taking the plunge, then this might be the best way in. Either it will satisfy you, or it will make your eventual jump to Cohen a little less daunting.

Statistical Power Analysis
Kevin R. Murphy and Brett Myors
1998 Lawrence Erlbaum Associates
0805829466

By far the slightest of the three books we recommend. But it's on our approved list because it is very clearly written, and while some ability to handle equations is required to make much use of the other two, this is written in such a way that those with very little confidence in mathematics could make some use of it. You could read it from start to finish in a day, but for all that, it is not insubstantial, and for some readers will provide all the power analysis they need.

In very large-scale ecological experiments, for example where an entire lake or river system must be manipulated, then replication may not be feasible. However, before arguing that you don't need replication in such experiments, think carefully. To help in this, we recommend reading Oksanen, L. (2001) Logic of experiments in ecology: is pseudoreplication a pseudoissue? *Oikos* **94**, 27–38.

There are a lot of software packages (many available free on the internet) that will do power calculations for you. Simply type 'statistical power calculations' into your favourite search engine. A particularly popular package can be found at http://calculators.stat.ucla.edu/powercalc/. An older but still popular one is G*Power available free from http://www.psycho.uni-duesseldorf.de/aap/projects/gpower/.

For more thoughts on using unbalanced designs, see Ruxton, G. D. (1998) Experimental design: minimising suffering may not always mean minimising number of subjects. *Animal Behaviour* **56**, 511–12.

Randomization

The paper that brought the idea of pseudoreplication to a wide audience was: Hurlbert, S. H. (1984) Pseudoreplication and the design of ecological field experiments. *Ecological Monographs* **54**, 187–211. This paper can be recommended as a good read with lots to say on design generally as well as a very clear discussion on pseudoreplication. All that and it discusses human sacrifices and presents an interesting historical insight into the lives of famous statisticians. Highly recommended: 2000 citations in the scientific literature must also be some kind of recommendation!

For further reading on the techniques required for effective comparisons between species, the following paper provides an intuitive introduction: Harvey, P. H. and Purvis, A. (1991) Comparative methods for explaining adaptations. *Nature* **351**, 619–24.

Your first book on statistics

Choosing and Using Statistics
Calvin Dytham
1998 Blackwell Science
0865426538

Those who like this book absolutely love it. Its great strength is that it can tell you very quickly what statistical test you want to do with your data and how to actually do that test on numerous different computer packages. This is a book about using statistics, not about the philosophy of statistical thinking.

Practical Statistics for Field Biology
Jim Fowler, Lou Cohen and Phil Jarvis
1998 John Wiley & Sons Ltd
0471982962

The opposite extreme to Dytham's book. There is no discussion on the details of specific computer packages and plenty on the underlying logic behind different statistical approaches. Don't get the impression that this is an esoteric book for maths-heads though, it is full of common sense advice and relies more on verbal reasoning than mathematics. Packed with illuminating and interesting examples.

Statistics for Biologists
R. C. Campbell
1989 (3rd edn) Cambridge University Press
0521369320

Perhaps a good compromise between the books above, since it does use specific computer packages (SPSS, Minitab and Genstat) to illuminate some of its examples, but also devotes space to discussion of the logic behind different types of analysis. It's a big book and the design is not flashy, but it covers a little more than either of the above, and does it in a clear and attractive style.

Practical Statistics for Experimental Biologists
Alastair C. Wardlaw
2000 John Wiley & Sons Ltd
0471988227

Very similar to Fowler and friends. The key differences are that it gives extensive examples developed for a specific computer package (Minitab) and the examples come more from the laboratory than the field.

Biomeasurement
Dawn Hawkins
2005 Oxford University Press
0199265151

Covers very similar ground to the preceding books and provides practical guidance in actually doing statistical tests in SPSS. Of all the books, this one is probably the most sympathetic to students who feel particularly nervous about statistics. However, this should not be taken as an indication that the material is dumbed down. Indeed a particular strength of this book is that it repeatedly makes use of real examples from the primary literature. Another strength is a particularly comprehensive companion website.

Asking Questions in Biology
Chris Barnard, Francis Gilbert and Peter McGregor
2001 (2nd edn) Prentice Hall
0130903701

This is a lot more than an elementary statistics textbook. Rather, it looks to take students through all the various processes of scientific enquiry: from asking the right question, to turning the question into a well-designed experiment, analysing the data from that experiment and finally communicating the outcome of the work. In 170 pages, this book cannot attempt to be comprehensive in any of the aspects it covers: but for those just starting to do their own experiments, this book is full of practical advice and wisdom.

Modern Statistics for the Life Sciences
Alan Grafen and Rosie Hails
2002 Oxford University Press
0199252319

The word 'modern' in the title should alert you that this book takes a rather different approach to traditional elementary statistical texts. Rather than introduce a number of statistical tests as discrete tools, it suggests an overarching framework based on construction of statistical models and the powerful and flexible modern technique of the General Linear Model. Our guess is that some day most statistics courses will be taught this way; in the mean time the traditional method still dominates. This book will appeal more to the student who wants to develop intuitive insight into the underlying logic of statistical testing.

More advanced (and expensive) general statistical books

Biometry
Robert R. Sokal and F. James Rohlf
1994 (3rd edn) W.H. Freeman
0716724111

For the issues it covers, Sokal and Rohlf is definitive. It does not cut corners, shirk from difficult issues, or encourage you to be anything other than very, very careful in your analysis. If your experiment has features that make your situation a little non-standard, and your elementary textbook doesn't cover the problem, then almost certainly these guys will. Its not an easy read, but if you put in the time it can deliver solutions that you'll struggle to find in easier books.

Biometrical Analysis
J. H. Zar
1998 (4th edn) Prentice Hall
013081542X

Some find Sokal and Rohlf a little stern; if so, Zar is the answer. Compared to elementary books, there is a lot more maths, but the book does take its time to explain clearly as it goes along. Highly recommended.

Experimental Design and Data Analysis for Biologists
Gerry P. Quinn and Michael J. Keough
2002 Cambridge University Press
0521009766

Less exhaustive in its treatment than either of the above, but this allows it to cover ground that they do not, including a considerable treatment of multivariate statistics. It also covers more experimental design than the other two. It has a philosophical tone, that encourages reading a large chunk at one sitting. However, it's also a pretty handy book in which to quickly look up a formula. This should be a real competitor for Zar. Try and have a look at both before deciding whose style you like best.

Experiments in Ecology
A. J. Underwood
1996 Cambridge University Press
0521556961

A blend of design and statistics, presented in an unusual way that will be perfect for some but not for all. Look at this book particularly if ANOVA is the key statistical method you are using. Good on subsampling too.

Ecology, Ethology and Animal Behaviour

Collecting data can be very challenging in these fields. Fortunately, there are some wonderful books to help you do this well.

Measuring Behaviour: An Introductory Guide
Paul Martin and Patrick Bateson
1993 (2nd edn) Cambridge University Press
0521446147

Many professors will have a well-thumbed copy of this on their shelves, that they bought as an undergraduate, but still find the need to consult. Brief, clear and packed with good advice—cannot be recommended highly enough.

Ecological Methodology
Charles J. Krebs
1999 (2nd edn) Addison-Welsey
0321021738

Much more substantial, and positively encyclopaedic on the design and analysis of ecological data collection. Beautifully written and presented; often funny and very wise.

Handbook of Ethological Methods
Philip N. Lehmer
1996 (2nd edn) Cambridge University Press
0521637503

Similar in many ways to the Krebs book in what it seeks to cover. Has a stronger emphasis on the mechanics of different pieces of equipment and good advice on fieldcraft, but less emphasis on analysis of the data after it is collected.

Index

3 2183 01564 6470